Annals of Mathematics Studies

Number 144

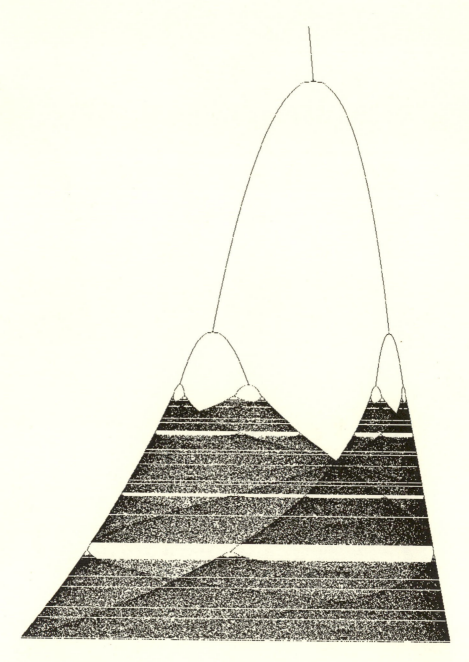

The orbit diagram of $f_a = ax(1-x)$ with $2.9 \leq a \leq 4$. The white "windows" correspond to hyperbolic dynamics when f_a has an attracting periodic orbit. The real Fatou Conjecture asserts that the windows are dense in the parametric interval.

The Real Fatou Conjecture

by

Jacek Graczyk and Grzegorz Świątek

PRINCETON UNIVERSITY PRESS

———————

PRINCETON, NEW JERSEY

1998

The Annals of Mathematics Studies are edited by
Luis A. Caffarelli, John N. Mather, and Elias M. Stein

Princeton University Press books are printed on acid-free paper and meet the
guidelines for permanence and durability of the Committee on Production
Guidelines for Book Longevity of the Council on Library Resources

Printed in the United States of America

10 9 8 7 6 5 4 3 2 1

Library of Congress Cataloging-in-Publication Data

Graczyk, Jacek.
The real Fatou conjecture / by Jacek Graczyk and Grzegorz Świątek.
p. cm. — (Annals of mathematics studies : 144)
Includes bibliographical references and index.
ISBN 0-691-00257-6 (cloth : alk. paper). — ISBN 0-691-00258-4
(pbk: alk. paper)
1. Geodesics (Mathematics) 2. Polynomials. 3. Mappings (Mathematics)
I. Świątek, Grzegorz, 1964– . II. Title. III. Series.
QA614.58.G73 1998
516.3'62—dc21   98-24386

The publisher would like to acknowledge the authors of this volume for
providing the camera-ready copy from which this book was printed

http://pup.princeton.edu

To the memory of Janos Bolyai

Contents

The Real Fatou Conjecture

Chapter 1

Review of Concepts

1.1 Theory of Quadratic Polynomials

Quadratic polynomials from the perspective of dynamical systems. Among non-linear smooth dynamical systems quadratic polynomials are analytically the simplest. Yet, far from being trivial, they have been subject of intense research for a couple of decades. A number of difficult papers have been produced and many key questions remain unsolved. Admittedly, some phenomena that are a staple of dynamical systems, such as homoclinic intersections, are impossible in one dimension. The flip side is that the simplicity of the system makes it possible to approach rigorously phenomena that are out of reach in higher dimensions, to just name the transition to chaos. For one reason or another, a number of mathematicians became interested in the very narrow field of quadratic polynomials.

Iteration of a quadratic polynomial leads to polynomials of progressively higher degrees and here the transparent simplicity of the system is lost. Given a polynomial of degree 2^{100}, how does one tell that it is an iteration of a quadratic; if so how can one exploit this fact dynamically? In real dynamics, a property of quadratic polynomials which is inherited under iteration is negative Schwarzian derivative. An impressive technique has been developed based on this property, see [30]. However, specific properties of quadratic polynomials and their iterations become more evident if they are viewed as mappings of the complex plane. The classical Julia-Fatou theory provides new insights. A powerful new tool known as quasiconformal deformations

3

becomes available. If two maps (say polynomials) are quasiconformally conjugated, one can perturb the conjugacy in such a way that a holomorphic family of conjugated systems of the same type (polynomials of the same degree) interpolating between the original ones is formed, see [39]. Nothing like this exists in the real theory. A polynomial can be perturbed explicitly by changing a parameter, but trying to manipulate the conjugacy between two real polynomials will lead to more complicated transformations, usually no more than continuous. And so from the mid-eighties on an idea of treating jointly the real and complex one-dimensional systems (see [39]) became increasingly popular. Real polynomials are right on the borderline and naturally became the proving ground for this concept.

The divide between real and complex dynamics. However, the merging of real and complex dynamics also encountered serious hurdles. The methods and the style of papers in both fields are different. In interval dynamics proofs are mostly long sequences of inequalities. To check a proof, one goes through all the inequalities and an occasional combinatorial lemma. The holomorphic dynamics is made of different ingredients. In many papers, there are few inequalities or formulas to go by. The proofs are made of concepts, often quite geometric in nature. For a non-specialist, checking a proof may present a formidable difficulty, since the key concepts are not easily put down as definitions or theorems.

A proof of Fatou's conjecture for real quadratic polynomials relies both on real and complex methods. However, the gist of many technical arguments is shifted from the real line to the complex plane. The relation between real and complex methods deserves to be carefully explained. This does not mean that we attempt to develop philosophical principles or heuristic arguments which even if widely accepted remain beyond the domain of mathematical proof. We simply try to formulate this relation rigorously.

The content of this book. In this book, our ambition is to present the proofs in a rigorous way accessible to the wide audience in dynamical systems and beyond. Hence, it is not to present all that is known about quadratic polynomials. For that, the most comprehensive source remains [7]. We skipped the complex case and concentrate on the proof of Fatou's conjecture for real quadratic polynomials. Such a limited approach gives our work a good logical structure, allows the presentation of a wide array of concepts, and best serves our goal of making a rigorous presentation.

1.1.1 Weak hyperbolicity of quadratic polynomials

There are two properties of real quadratic polynomials that make this proof work. The first is known as "complex bounds" of renormalization. A quadratic polynomial is generally not expanding, but it always stretches sets in the large scale. The meaning of this "large scale" expansion is explained in Douady and Hubbard's definition of a polynomial-like map (see [8]): namely that the domain of the mapping, assumed to be a topological disk is mapped on a strictly larger region, which contains the closure of the original domain. The "strength" of this expansion can be measured by the width of the set-theoretical difference between the range and the domain. Renormalization of real unimodal maps is a phenomenon when an interval (a so-called restrictive interval) is mapped into itself by an iterate and this transformation is unimodal. If the original system was a quadratic polynomial, this first return map is a polynomial of high degree. The complex bounds property says that if a topological disk is suitably chosen around the restrictive interval, then the first return map becomes polynomial-like, with only one critical point in its domain, and with "strength" bounded away from 0.

The second property is related to the concept of inducing. In 1981 Michael Jakobson proved the existence of invariant measures for a large set of unimodal maps. His method was based on replacing the original mapping on pieces of the domain by iterations. In the end, he obtained a map defined almost everywhere, with infinitely many branches each being an iterate of the original transformation, all monotone, expanding and mapping onto a fixed interval. Later research showed that this property was quite prevalent. However, it cannot hold for infinitely renormalizable mappings for topological reasons. Nevertheless, even for those it remains true that high iterations become expanding, if chosen appropriately. To fully exploit this phenomenon, in 1993 we introduced a class of so-called box mappings, see [13]. In the language of box mappings, the property becomes the increase of certain conformal moduli.

It should be emphasized that the only case when both properties are satisfied is the real quadratic family. The first property belongs to real systems and is true for unimodal polynomials of any degree with generalizations to real-analytic mappings, see [26]. However, it has no counterpart for complex quadratic polynomials, see [34]. The second property is not applicable in general if the degree is greater than 2.

1.2   Dense Hyperbolicity

1.2.1   Theorem and its consequences

The Dense Hyperbolicity Theorem.   *In the real quadratic family*

$$f_a(x) = ax(1 - x) \, , \, 0 < a \leq 4$$

the mapping f_a has an attracting cycle, and thus is hyperbolic, for an open and dense set of parameters a.

The Dense Hyperbolicity Theorem follows from the Main Theorem which gives an analytically checkable condition for instability in the real quadratic family.

Main Theorem. *Let f and $\hat{f}$ be two real quadratic polynomials with a bounded forward critical orbits and no attracting or indifferent cycles. Then, if they are topologically conjugate, the conjugacy extends to a quasiconformal conjugacy between their analytic continuations to the complex plane.*

Derivation of the Dense Hyperbolicity Theorem.   We show that the Main Theorem implies the Dense Hyperbolicity Theorem. The reduction is based on three facts, two from complex dynamics and one from real, which we state here with proofs.

Fact 1.2.1 *Consider the set of quadratic polynomials $f_a(z) = az(1 - z)$ where a is analytic parameter. For some a, let $C_a \subset C$ denote the set of all b such that f_a and f_b are conjugated on the complex plane by a quasiconformal homeomorphism. Then C_a is either $\{a\}$, or is open.*

Proof:
The proof of this fact follows by the method of quasiconformal deformations introduced in [38]. The result is implicit in [27]. Suppose that f_a is q.c. conjugate with f_b, and $a \neq b$. The conjugacy H can not be conformal on the whole plane and thus there is an f_b-invariant Beltrami coefficient μ which is not identically equal to 0. We will show that b belongs to C_a together with an open ball. To this end observe that for every $c \in \mathbb{C}$ such that $|c| < 1/||\mu||_\infty$, $c \cdot \mu$ is an f_b-invariant Beltrami coefficient. Let H_c be a solution of the Beltrami equation

$$\frac{dH}{d\bar{z}} = c\mu \, \frac{dH}{dz}$$

normalized by the condition that $0, 1, \infty$ are the fixed points. By the measurable Riemann mapping theorem, see [3], H_c depends analytically on c and $f_{r(c)} = H_c \circ f_b \circ H_c^{-1}$ is an analytic family of analytic functions. By topology, $f_{r(c)}$ is a $2-1$ branched covering of the Riemann sphere which fixes 0 and ∞. Since ∞ is also a branching point, $f_{r(c)}$ is a family of quadratic polynomials. The eigenvalue $e(c) := \frac{d}{dz} f_{r(c)}(0)$ is an analytic function of c and the image of the ball $|c| < 1/||d\mu||_\infty$ by e is either a point or an open set. The first possibility is excluded since $r(1) = a$ and $r(0) = b$. The function $e(c)$ gives an analytic reparametrization $e(c)z(1-z)$ of $f_{r(c)}$.

$\square$

Fact 1.2.2 *There are only countably many complex values of a for which the map $a \to ax(1-x)$ has a neutral periodic point.*

Proof:
We will prove a stronger statement.
If $k > 0$, and λ is a complex number with absolute value less or equal to 1, then the pair of equations

$$f_a^k(z) = z \quad and \quad \frac{df_a^k}{dz}(z) = \lambda$$

has only finitely many solutions (a, z).
  The proof is based on the following theorem about Riemann surfaces of algebraic functions, see [9] Theorem IV.II.4 on pages 231-232,

Fact 1.2.3 *Consider the equation $P(a, z) = 0$ where P is an irreducible polynomial of two complex variables. Then the set of solutions, compactified by adding points at infinity, has the structure of a compact Riemann surface. Moreover, projections on a and z are meromorphic of this surface.*

This theorem applied to the polynomial $f_a^k(z) - z = 0$ implies that the set of solutions splits into the union of finitely many compact Riemann surfaces. On each of these, the function

$$\frac{df_a^k}{dz}(z)$$

is meromorphic. If it takes value λ infinitely many times, by the identity principle it must be constant on one of the surfaces, call it S. If a pair (a, z) solves both equations, it means that a must be in the connectedness locus in the parameter space, and z is in the filled Julia set. Hence, both projections map the finite points of S into a bounded set in the complex plane. The image of S under either projection must be compact, since the projection is continuous. But since the projections are also open mappings or constant, the image of either of them is just a point. Hence, S must be a point, which is impossible.

$\square$

Fact 1.2.4 *Two real quadratic polynomials f and g with bounded critical orbits and such that f has no attracting periodic orbit, normalized so that their critical points z_f and z_g, respectively, are maxima, are topologically conjugate on the real line if and only if for every $n > 0$ both differences $f^n(z_f) - z_f$ and $g^n(z_g) - z_g$ have the same sign.*

Proof:
The "only if" part is obvious and the rest is contained in Theorem 2.10 in [17].

$\square$

Suppose that $f_a(x) = ax(1-x)$ has no stable periodic orbits. Let T_a denote the set of parameter values b such that $f_b(x) = bx(1-x)$

is topologically conjugate to f_a on the real line. Then T_a is closed. Indeed, if a sequence $b_n \in T_a$ converges to b, the signs of $f_{b_n}^k(1/2) - 1/2$ remain fixed for all k. By continuity, the differences $f_b^k(1/2) - 1/2$ either remain of the same sign, in which case our assertion follows from Fact 1.2.4, or some may vanish. If one of them vanishes, it means that $1/2$ is periodic by f_b. But then the implicit function theorem implies that for all parameters in a neighborhood of b there is a stable periodic orbit. This is a contradiction since f_{b_n} for all n have no such orbits.

Now Fact 1.2.1 means that for any $a \in (0, 4]$ such that f_a has only repelling periodic orbits, the quasiconformal class C_a intersected with the real line is either a point or is open. The Main Theorem means that $C_a \cap \mathbb{R} = T_a$. Since T_a is closed, it must be a point.

To prove the Dense Hyperbolicity Theorem, let $a \in (0, 4]$. In view of Fact 1.2.2, we can assume without loss of generality that f_a has only repelling periodic orbits. In every neighborhood of a we can find $a_1 \neq a$ so that f_{a_1} still only has repelling periodic orbits. As observed in the preceding paragraph, f_a and f_{a_1} are not topologically conjugate. In view of Fact 1.2.4 this means that for some k the differences $f_a^k(1/2) - 1/2$ and $f_{a_1}^k(1/2) - 1/2$ have different signs. By the intermediate value theorem, for some a_0 between a_1 and a we get $f_{a_0}^k(1/2) = 1/2$. This means that $1/2$ belongs to an attracting periodic orbit and proves the Dense Hyperbolicity Theorem.

Historical notes. The Dense Hyperbolicity Conjecture has had a long history. In a paper from 1920, see [10], Fatou expressed the belief that "general" (generic in today's language?) rational maps are expanding on the Julia set. Our result may be regarded as progress in the verification of his conjecture. More recently, the fundamental work of Milnor and Thurston, see [32], showed the monotonicity of the kneading invariant in the quadratic family. They also conjectured that the set of parameter values for which attractive periodic orbits exist is dense, which means that the kneading sequence is strictly increasing unless it is periodic. The Dense Hyperbolicity Theorem implies Milnor and Thurston's conjecture. Otherwise, we would have an interval in the parameter space filled with polynomials with an aperiodic kneading sequence, in violation of the Dense Hyperbolicity Theorem.

Yoccoz, [41], proved that a non-hyperbolic quadratic polynomial with a fixed non-periodic kneading sequence is unique up to an affine conjugacy unless it is infinitely renormalizable. This implied our Main Theorem in all cases except for the infinitely renormalizable. His method is different from one explained in this book and the Main Theorem does not even appear as a step in the proof. Instead, geometric estimates are established in the phase space and then used in the parameter space to explicitly show that the set T_a is a point.

In the infinitely renormalizable case, the work of [39] proved the Main Theorem for infinitely renormalizable polynomials of bounded combinatorial type. The paper [24] achieved the same for some infinitely renormalizable quadratic polynomials not covered by [39]. This book follows the method of [16].

1.2.2   Reduced theorem

All unimodal maps that are mentioned in this book are normalized so that $f : [-1, 1] \to [-1, 1]$, $f(x) = f(-x)$, 0 is the local maximum and $f' \neq 0$ elsewhere. Let us adopt the following terminology.

Definition 1.2.1 *A unimodal mapping f, with the critical point at 0, is called* critically recurrent *provided that:*

- *f has no attracting or indifferent periodic cycles,*

- *$0 \in \overline{\{f^n(0) : n > 0\}}$.*

The Main Theorem can be easily derived from a purely real statement.

Reduced Theorem. *Let f and $\hat{f}$ be topologically conjugate. Suppose that both are critically recurrent real quadratic polynomials. Then there is a quasi-symmetric orientation-preserving homeomorphism h of the interval $(-1, 1)$ onto itself which satisfies*

$$\hat{f}^i(h(0)) = h(f^i(0))$$

for every positive integer i.

In other words, the Reduced Theorem asserts that there exists a quasi-symmetric conjugacy on forward critical orbits of our two polynomials.

Reduced Theorem implies Main Theorem. In the statement of the Reduced Theorem, we confine ourselves to critically recurrent maps. In fact, the Reduced Theorem remains true if the assumption of critical recurrence is replaced with all periodic orbits being repelling. The case in between, when all critical orbits are repelling, but the critical orbit is not recurrent, is known to the experts as the *Misiurewicz case*. The Main Theorem in the Misiurewicz case follows from Theorem 1 in [22] (see also Fact 2.2.1).

Let us now assume the Reduced Theorem, including the Misiurewicz case. That is, given two polynomials f and $\hat{f}$ which satisfy the hypotheses of the Main Theorem, we have a quasi-symmetric mapping h from the interval $(-1, 1)$ onto itself which satisfies

$$h(f^k(0)) = \hat{f}^k(0)$$

for all positive integers k.

We proceed to derive the Main Theorem. The argument we present follows [39]. First, we build a quasiconformal homeomorphism $H_0 : \mathbb{C} \to \mathbb{C}$ which satisfies the following:

- $H_0(f^k(0)) = \hat{f}^k(0)$ for all positive integers k,

- outside of $D(0, 100)$, H_0 is the Böttker coordinate in the basin of infinity.

That such H_0 exists is clear, since one can construct it on a neighborhood U of the interval $[-1, 1]$ in the complex plane from Beurling-Ahlfors' extension theorem applied to h and then interpolate between U and the domain of the Böttker coordinate by a quasiconformal map. That such an interpolation is possible follows formally from Lemma 5.3.1 proved later in our book. We proceed inductively. Given H_i for which the first of the two aforementioned properties is satisfied, we choose H_{i+1} to be the pull-back of H_i, that is

$$H_i \circ f = \hat{f} \circ H_{i+1}$$

and H_{i+1} is chosen to preserve the orientation on the interval $(-1, 1)$.

The first condition is satisfied again by H_{i+1}. We know that $H_{i+1}(f^k(0))$ is either $\hat{f}^k(0)$ or $-\hat{f}^k(0)$. By the requirement that the ordering is preserved, $H_{i+1}(f^k(0))$ and $\hat{f}^k(0)$ are on the same side of

0, and since the kneading sequences of f and $\hat{f}$ are the same, then $\hat{f}^k(0)$ remains as the only choice.

We also observe by induction that H_i coincides with the Böttker coordinate on the set

$$A_i := f^{-i}(\mathbb{C} \setminus D(0, 100))$$

and that $H_i, H_{i+1}, \cdots$ are all the same on A_i. Since the union of A_i is the complement of the Julia set of f, hence is dense in the complex plane, we see that if the sequence H_i converges in the C^0 topology on the complex plane to a homeomorphism H_∞, then H_∞ conjugates between f and $\hat{f}$. Since all H_i are quasiconformal with the same bound on the maximal dilatation, then family is equicontinuous on compact subsets and the limit of any convergent subsequence is quasiconformal. So it is enough to show that if H_{i_j} and $H_{i_{j'}}$ are two subsequences convergent almost uniformly in the plane, then the limits are equal on the complement of the Julia set. This is clear because the sequence H_i stabilizes on relatively compact sets in the complement of the Julia set.

This proves the Main Theorem.

1.3   Steps of the Proof of Dense Hyperbolicity

In this section we endeavor to show the main stages of the proof of the Reduced Theorem, at the same time introducing the main concepts. Let us suppose that we are in the situation of the Reduced Theorem.

1.3.1   Regularly returning sets and box mappings

Consider a quadratic polynomial f, viewed either as a real or complex map. Following the concept of a "nice set" of [28], we will call an open set U which contains the critical point *regularly returning* if $f^n(\partial U)$ is disjoint from U for all positive n. Then the first return map on U has useful properties which can be summarized as follows.

Let us recall the concepts of the first return and first entry times. If $F : X \to X'$ is a mapping of a set X into a set $X' \supset X$, and $X' \supset Y$, then the *first entry time* of F into Y is

$$e(x) := \inf\{k = 0, 1, \cdots : F^k(x) \in Y\}$$

for $x \in X$. Note that $F^k(x)$ may not be defined for all k. If this set is empty, then following the usual convention $e(x) = \infty$. The *first return time* from Y into itself is

$$r(x) := \inf\{k = 1, \cdots : F^k(x) \in Y\}$$

defined for $x \in Y$. The mappings $x \to F^{e(x)}(x)$ and $x \to F^{r(x)}(x)$ are called the first entry and first return maps, respectively, and defined wherever the corresponding time is finite. Clearly, the first return map from Y into itself is the same as F restricted to $Y \cap X$ followed by the first entry map. The first entry map is always defined and equal to the identity on Y.

Fact 1.3.1 *Let f be a quadratic polynomial, viewed as a real or complex map, and U a regularly returning topological disk (interval in the real case). Then the first entry map ϕ from U into itself is defined on a disjoint union of disks, ϕ restricted to each connected component of its domain is one-to-one, an iterate of f, and maps onto U.*

Proof:
Let x belong to the domain of ϕ. Recall that $e(x) < \infty$ is the first entry time of x into U. Let V be the connected component containing x of the set $f^{-e(x)}(U)$. Observe that for all points $y \in V$, $y, f(y), \cdots, f^{e(x)-1}(y) \notin \overline{U}$. Otherwise, for some y and intermediate image would be in ∂U, and then it could not return into U by the assumption that U is regularly returning. On the other hand, $f^{e(x)}(\partial V) \subset \partial U$ by continuity, and then these points never enter U, again by the property of regular returning. Hence, ϕ is undefined on ∂V and V is the connected component of the domain of ϕ which contains x. Also, $\phi(y) = f^{e(x)}(y)$ on V, so indeed ϕ restricted to V is an iterate of f.

From now, think of ϕ as being restricted to V. Since $f^{e(x)}(\partial V) \subset \partial U$, the map ϕ is proper, that is, preimages of compact sets relative to U are compact relative to V. Also, ϕ has no critical points because the only critical point of f is in U, and ϕ is the first entry map. If f is real, it is already obvious that ϕ maps V monotonely onto U. In the complex situation, the fact that ϕ restricted to V is proper implies that ϕ is a covering of U by V and hence V is simply connected. A holomorphic map between two conformal disks whose derivative is never 0 must be univalent by the reflection principle.

$\square$

Box mappings. The definition we give below applies to both real and complex mappings, once we remember that an interval is a topological disk on the real line.

Definition 1.3.1 *Given a topological disk B' containing a point c_0, a type II box mapping ϕ is an analytic map from some open set W into B' which satisfies the following:*

- *$W \cap \partial B' = \emptyset$,*

- *all connected components of W are topological disks, and ϕ is proper from any component of W into B',*

- *ϕ has exactly one critical point, called c_0, which must be of order 2, and the connected component of W which contains c_0 is typically denoted with B,*

- *$\overline{B} \subset B'$.*

In the complex case, if B' and all connected components of W are Jordan domains, we will call ϕ a *Jordan* type II box map. Strictly speaking, the extra condition is not needed in this paper, but it is sufficient to consider Jordan box mappings. So we will tacitly assume that. Restrictions of a box map to connected components of its domain will be referred to as *branches* of ϕ. In particular, $\phi_{|B}$ is called the *central branch*, and typically denoted with ψ. Notice that in the complex case all branches of ϕ map onto B'.

The following is a prototype of a type II box mapping. Consider a regularly returning topological disk U for some quadratic polynomial. If one makes ϕ the first return map from U into itself on U, and the the first entry map into U elsewhere, from Fact 1.3.1 we see that ϕ is a type II box mapping provided that the critical orbit eventually returns to U and the central branch is compactly contained in U. Also, this box map is Jordan provided that U is a Jordan domain.

The inducing algorithm. Given a type II box mapping ϕ with range B' and central domain B, we can construct another type II induced map $\tilde{\phi}$ which on B is the first return map of ϕ from the central domain B into itself and elsewhere the first entry map of ϕ into B. With respect to the ambient dynamics of f, B is a regularly returning topological disk and $\tilde{\phi}$ can also be viewed as the first return and first entry map into B in terms of f. We will call the operation of passing from ϕ to $\tilde{\phi}$ a *simple inducing step*. The *inducing algorithm* simply refers to the iteration of this process. Thus, from a type II box mapping ϕ_0, we derive a sequence of type II box mappings ϕ_n so that ϕ_n for $n > 0$ is derived from ϕ_{n-1} by a simple inducing step. This process might terminate if ϕ_n has no central domain, that is, if the critical value of ϕ_{n-1} never returns to the central domain of ϕ_{n-1}. This situation is ruled out if the ambient map f is critically recurrent. In the further discussion we assume that the critical value always returns.

Filling-in and critical filling. For a type II box mapping, the first return into its central domain B can be decomposed into two steps. Consider ϕ' the first entry map of ϕ into B. Then we can define a map φ which is the same as ϕ on B and the same as ϕ' elsewhere. The map φ is an example of a *type I box mapping*. It is possible to write down an abstract definition of a type I box mapping similar to Definition 1.3.1. However, we will only consider type I box mappings induced by type II maps in the way described in this paragraph.

The passage from ϕ to φ is called *filling-in*. Then $\tilde{\phi}$, of type II, is the same as φ outside of B and is given by $\phi' \circ \varphi$ on B. This second step of passing from φ to $\tilde{\phi}$ is called *critical filling*. In the future, the term *box mapping* will refer to both type I and type II box mappings. A *simple inducing step* is also defined for type I box mappings: it is a critical filling followed by a filling-in.

Close returns and type I inducing step. We say that a box mapping shows a *close return* if the critical value is in the central domain, otherwise the return is described as *non-close*.

Definition 1.3.2 *The escaping time of ϕ is defined to be*

$$E := \min\{i = 1, 2, 3, \ldots : \psi^i(c_0) \notin B\}$$

By convention, $E = \infty$ if the above set is empty.

Close and non-close returns are distinguished in terms of the escaping time. Namely, $E = 1$ if and only if ϕ shows a non-close return.

Now we define a *type I inducing step*. Let ϕ be a type I box mapping with escaping time $E < \infty$. The type I inducing step is defined as the E-th iteration of the simple inducing step. In other words, we continue simple inducing steps ($E - 1$ times) until the first non-close return occurs and then proceed with one more simple inducing step. The type I inducing step transforms a type I box mapping into another type I box mapping.

1.3.2   Quadratic-like returns

Definition 1.3.3 *A box mapping is called* terminal *if there is an open interval $I \subset B$ containing the critical point of ϕ so that $\phi(I) \subset I$ and $\phi(\partial I) \subset \partial I$. The interval I (which must be unique) will then be called the* restrictive interval *of ϕ.*

Clearly, the critical point of a terminal box mapping has the escaping time from the central domain $E = \infty$. The converse is also true. Indeed, let ψ denote the central branch of ϕ, and B its central domain. If $E = \infty$, then the critical value must be contained in $\psi^{-n}(B)$ for any $n \geq 0$. These intervals form a descending sequence, and the intersection must be more than a point, since otherwise 0 would be fixed by ψ. So the intersection is an non-degenerate interval symmetric with respect to 0 and invariant under ψ which meets the criterion of Definition 1.3.3.

On the level of ambient dynamics, a terminal map corresponds to some *restrictive interval*.

Definition 1.3.4 *Suppose that f is a unimodal mapping. An open interval J, smaller than $(-1, 1)$, and symmetric with respect to 0 is called a* restrictive interval *for f if the first return time into J is the same for all points of J.*

Observe that $J, f(J), \cdots, f^{n-1}(J)$ are disjoint, if the first return time into J is n. Then f^n restricted to J is the quadratic map composed with a diffeomorphism onto $f^n(J)$. Of particular importance are *locally maximal* restrictive intervals. Notice that f^n maps the boundary of J into the boundary of J, for otherwise J could be extended and still remain a restrictive interval. Now consider the linear map A that transforms J onto $[-1, 1]$. Then, if J is a locally maximal restrictive interval, the mapping

$$g = A \circ f^n \circ A^{-1}$$

is a unimodal mapping. We will call g a *renormalization* of f. Since it is only required that J be locally maximal, f can have many renormalizations, but one of them is distinguished: namely the one corresponding to the maximal restrictive interval smaller than $[-1, 1]$. That one will be called the *first renormalization* of f.

For us, the important observation is:

Fact 1.3.2 *The restrictive interval of a terminal box mapping induced by f (see Definition 1.3.3) is also a restrictive interval of f itself.*

The polynomial-like property.

Theorem 1.1 *Let f be a renormalizable quadratic polynomial without attracting or indifferent periodic orbits and let $I_1 \supset \cdots$ be the sequence, finite or not, of its locally maximal restrictive intervals (see Definition 1.3.4). Let f_i denote the first return map into I_i. Then, for every i f_i is conjugate to a unimodal quadratic polynomial by an L-quasi-symmetric homeomorphism sending I_i to $(-1, 1)$. The constant L is independent of f.*

Theorem 1.1 is a major improvement over earlier versions of proof of the Real Fatou Conjecture. It has two independent proofs, in [15] and [26]. It follows from the fact that every f_i has an analytic continuation to a quadratic-like mapping in the sense of Douady and Hubbard and so that the modulus of the nesting of the domain of this mapping inside its range is bounded from below by a positive constant, see [8], also [30] for the "uniform" version. In other words, all renormalizations of real quadratic polynomials are quadratic-like with uniform bounds.

1.3.3   Initial construction and geometry of inducing

Initial induced map. Let f be a unimodal critically recurrent map. One can build a real mapping Φ *induced* by f. The word induced means that the domain of the map is generally disconnected and on each connected component the mapping is an iterate of f. It is easy to understand Φ on the interval. A unimodal polynomial has a fixed point $q > 0$. The interval $(-q, q)$ is the fundamental domain of the dynamics since every orbit except for some eventually periodic ones passes through it infinitely many times. The control of these passages was the starting point of several constructions and is also our starting point. This interval is also regularly returning. The real map Φ is the first return map of f from $(-q, q)$ into itself and the first entry map into $(-q, q)$ elsewhere. It is a real type II box mapping.

Next, we go one step further to define a holomorphic mapping ϕ which is an analytic continuation of Φ. To this end, we must find a regularly returning topological disk in the complex plane which intersects the real line along $(-q, q)$. Such a disk, moreover a Jordan domain, exists as observed by Yoccoz. It is delimited by two pairs of rays from infinity, each pair converging at an endpoint of $(-q, q)$, and an equipotential curve. Details will be explained later, for now it is only essential that such a set can be found.

The linear growth of moduli. We begin with the initial type II holomorphic box mapping ϕ and first fill it in to get a type I map. Then we proceed by a sequence of type I inducing steps. Under our standing assumption that the mapping f is critically recurrent, the possibility of obtaining a terminal map is the only obstacle to performing a type I inducing step. This situation can indeed occur and leads to renormalization.

We will prove the following theorem:

Theorem 1.2 *Let ϕ be a type II holomorphic box mapping, and let B and B' denote the domain and range of its central branch, respectively. Let ϕ_0 be the box mapping obtained from ϕ by filling-in and ϕ_i form a sequence, finite or not, of holomorphic box mappings set up so that ϕ_{i+1} is derived from ϕ_i by the type I inducing step for $i \geq 0$. Suppose that for ϕ, $\mod (B' \setminus B) \geq \alpha_0$. For every $\alpha_0 > 0$*

there is a number $C > 0$ with the property that for every i

$$\text{mod}\,(B_i' \setminus B_i) \geq C \cdot i\,.$$

In particular, for every i we have that every connected component of the domain of ϕ_i contained in B_i' is separated from the complement of B_i' by an annulus with modulus at least $C \cdot i$.

Theorem 1.2 establishes a linear, but not necessarily monotone, growth of the moduli.

The intuitive meaning of the estimate of Theorem 1.2 is that the holomorphic box mappings simplify when induced. All univalent branches become close to linear and the folding branch after a proper rescaling becomes close to quadratic superexponentially fast. In all cases when induced box mappings work well as the subsequent approximates of the limit dynamics the property of linear growth of moduli shows that the limit dynamics is in fact trivial.

1.3.4 Branchwise equivalence

In the situation of the Reduced Theorem, the next step is to carry out the construction of the initial box mapping from the previous section for a pair of topologically conjugate quadratic polynomials f and $\hat{f}$. Clearly, the mapping $\hat{\Phi}$ will be topologically conjugate to Φ. It is less clear whether their holomorphic continuations ϕ_0 and $\hat{\phi}_0$ are topologically conjugate; this goes back to the question of whether f and $\hat{f}$ are conjugate as complex polynomials. We are not willing to assume that. We will mostly rely on a much weaker concept of *branchwise equivalence*.

Definition 1.3.5 *A homeomorphism Υ of the complex plane onto itself is considered a* branchwise equivalence *between two box mappings ϕ and $\hat{\phi}$ provided that Υ maps the domain of ϕ onto the domain of $\hat{\phi}$ and satisfies the following dynamical condition:*

- *If V is a connected component of the domain of ϕ, ζ is ϕ restricted to V, while $\hat{\zeta}$ is $\hat{\phi}$ restricted to $\Upsilon(V)$, then*

$$\hat{\zeta} \circ \Upsilon = \Upsilon \circ \zeta$$

on the boundary of V.

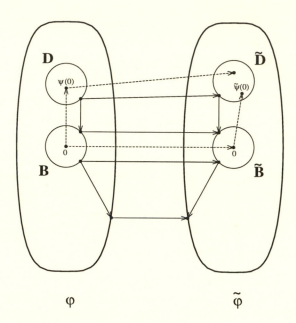

Figure 1.1: Branchwise equivalence between type I holomorphic box mappings ϕ and $\tilde{\phi}$. Diagrams commute for points on the boundary of the domain of ϕ, not necessarily for other points.

This definition assumes that ϕ extends from each connected component of its domain to the closure; this is clearly the case when ϕ is piecewise iterations of a polynomial. The notion of branchwise equivalence will only be used to box mappings defined in the complex plane, not just on the line.

A branchwise equivalence between type I holomorphic box mappings ϕ and $\hat{\phi}$ is shown on Figure 1.1.

The concept of branchwise equivalence is connected with the pull-back construction which will be introduced later. Unlike the topological conjugacy, a branchwise equivalence is far from unique. A rough classification of branchwise equivalences is given by the following similarity relation.

Similarity relation.

Definition 1.3.6 *Let ϕ and $\hat{\phi}$ be box mappings, real or holomorphic, with branchwise equivalences Υ_1 and Υ_2. Let D be the boundary of the domain of ϕ. We say that Υ_1 and Υ_2 are similar if they coincide on D and are homotopic to each other relative to D.*

We will use the notation $[\Upsilon]$ for the similarity class of Υ.

Statement of the initial construction. The essential features of the construction of the initial induced mappings are contained in the statement of the following theorem.

Theorem 1.3 *Suppose that f and $\hat{f}$ are unimodal mappings and quadratic polynomials. Moreover, assume that they are topologically conjugate and their critical orbits omit the fixed points 0 and q. In this situation, one can construct:*

- *A pair of type I box mappings ϕ_0 and $\hat{\phi}_0$, induced by f and $\hat{f}$, respectively. The restrictions of the domains of ϕ and $\hat{\phi}$ to the real line are dense in $(-1, 1)$. The sets of points which are forever iterated forward by univalent branches of ϕ_0 or $\hat{\phi}_0$ are totally disconnected.*

- *A branchwise equivalence Υ between ϕ_0 and $\hat{\phi}_0$, similar to a branchwise equivalence which is equal to the topological conjugacy between f and $\hat{f}$ on the real line.*

Furthermore, there are constants Q and $\epsilon > 0$ independent of f, so that

- Υ *is Q-quasiconformal and $\overline{\Upsilon(z)} = \Upsilon(\bar{z})$ for all $z \in \mathbb{C}$,*

- $\mathrm{mod}\,(B' \setminus B) \geq \epsilon$ *and* $\mathrm{mod}\,(\hat{B}' \setminus \hat{B}) \geq \epsilon$ *where B, B' and $\hat{B}$, $\hat{B}'$ are the domain and range of the central branch for ϕ_0 and $\hat{\phi}_0$, respectively.*

Theorem 1.3 does not appear in the literature, but it can hardly be regarded as new. The new element is the use of Yoccoz partitions. The most delicate part of the claim is the constant bound on the maximal dilatation of the branchwise equivalence. We give a modern proof based on the λ-lemma, but it could probably also be achieved by the tools of classical theory following Fatou. Similar results were obtained, for example, in [21]. A version on the real line, where the existence of a quasi-symmetric branchwise equivalence between Φ and $\hat{\Phi}$ is shown, comes from [24].

1.3.5   Pull-back

Suppose that two type I box mappings ϕ and $\hat{\phi}$ are given with a branchwise equivalence Υ between them. Suppose that this branchwise equivalence is *critically consistent*, i.e. $\Upsilon(\phi(0)) = \hat{\phi}(0)$ where 0 is the critical point of both box mappings. If ϕ_1 and $\hat{\phi}_1$ are obtained from ϕ and $\hat{\phi}$, respectively, by a simple inducing step, we can find a branchwise equivalence $\Pi(\Upsilon)$ between them as the limit of a sequence of maps Υ_n where $\Upsilon_0 := \Upsilon$ and for $n > 0$ the map Υ_n satisfies

$$\Upsilon_{n-1} \circ \phi = \hat{\phi} \circ \Upsilon_n$$

on the domain of ϕ and coincides with Υ_{n-1} elsewhere. The operation of finding such Υ_n is called the *pull-back* of Υ_{n-1}. We have already seen a similar construction in the proof that the Reduced Theorem implies the Main Theorem. Details of this construction will be described in Chapter 5. For now, we just assume that such a branchwise equivalence $\Pi(\Upsilon)$ is well-defined. Since we can always lift a homotopy by the pull-back, it is not surprising that Π preserves similarity classes. A similarity class of branchwise equivalences between ϕ and $\hat{\phi}$ is *critically consistent* if it contains a critically consistent representative. Hence, critically consistent similarity classes of branchwise equivalences between ϕ and $\hat{\phi}$ are mapped by Π into similarity classes of branchwise equivalences between ϕ_1 and $\hat{\phi}_1$.

Pull-back on quasiconformal representatives. We encounter the following technical problem. Suppose that a similarity class of branchwise equivalences contains a K-quasiconformal representative. Can we also claim that the similarity class derived by pull-back contains a quasiconformal representative with a specific dilatation bound K'?

Proposition 1 *Let ϕ and $\hat{\phi}$ be type I holomorphic box mappings. Suppose that under simple inducing steps ϕ shows $E-1$ close returns concluded with a non-close return in the E-th step. Let Υ be a K-quasiconformal branchwise equivalence between ϕ and $\hat{\phi}$. Assume that:*

- *the similarity class $[\Upsilon]$ is critically consistent,*

- *inductively, if ϕ_i and $\hat{\phi}_i$ are obtained from ϕ and $\hat{\phi}$, respectively, by i simple inducing steps for $i = 0, \cdots, E$, then $\Pi^i([\Upsilon])$ is a well-defined and critically consistent branchwise equivalence between them,*

- *if D is the domain of a branch of $\hat{\phi}$ and $D \subset \hat{B}'$, then $\mathrm{mod}\,(\hat{B}' \setminus D)$ is greater than ϵ.*

Then, for every $\epsilon > 0$ there is Q so that the similarity class of the branchwise equivalences $\Pi^E([\Upsilon])$ between ϕ_E and $\hat{\phi}_E$ has a QK-quasiconformal representative. Moreover, if $\epsilon \geq 4\log 8$, then $Q = \exp(Q' \exp(-\frac{\epsilon}{4}))$ where Q' is a constant, independent of all parameters.

Construction of branchwise equivalences. We begin with ϕ_0, $\hat{\phi}_0$ and the branchwise equivalence Υ derived from Theorem 1.3. Use filling-in once to pass to type I box mappings which are more natural for pull-back. Then we proceed to obtain mappings ϕ_n and $\hat{\phi}_n$ by iterating type I inducing steps, and we build branchwise equivalences Υ_n by pull-back. The branchwise equivalence Υ_n is chosen as the representative of the similarity class derived from $[\Upsilon_{n-1}]$ by pull-back with the smallest possible maximal dilatation bound. As a consequence of the fact that f and $\hat{f}$ are topologically conjugate on $[-1, 1]$, and that Υ was similar to a branchwise equivalence extending the conjugacy from the interval, we will keep getting critically consistent similarity classes $[\Upsilon_n]$. By each step, the quasiconformal bound of Υ_n is multiplied by Q given by Proposition 1. This process either goes on to infinity, or stops when a terminal mapping is obtained. The maximal dilatation of Υ_n turns out to be bounded by a constant based on the fact that the parameter ϵ from Proposition 1 grows at a uniform linear rate due to Theorem 1.2

That means that the product of corrections Q is bounded by a constant. Hence all branchwise equivalences Υ_n are uniformly quasiconformal. In the non-renormalizable case this immediately gives the Main Theorem, as the branchwise equivalences restricted to the real line tend to the conjugacy in the C^0 topology, and the limit is necessarily quasiconformal. In the renormalizable case, we will get a branchwise equivalence between the terminal box mappings. It should be emphasized that this branchwise equivalence is quasiconformal with a bound independent of the initial f and $\hat{f}$.

1.3.6 Conclusion of dense hyperbolicity

By Section 1.3, we have by now proved the Reduced Theorem. That is, given two polynomials f and $\hat{f}$ which satisfy the hypotheses of the Main Theorem, we have a quasi-symmetric mapping h from the interval $(-1, 1)$ onto itself which satisfies

$$h(f^k(0)) = \hat{f}^k(0)$$

for all positive integers k. The Main Theorem and the Dense Hyperbolicity Theorem follow as shown at the beginning of this section.

Chapter 2

Quasiconformal Gluing

The main objective of this chapter is to formulate Theorem 2.1 and conclude the Reduced Theorem. We will introduce a concept of *saturated maps* which facilitates gluing of quasiconformal branchwise equivalences. Given a pair of terminal box mappings we proceed by removing their central branches and replacing monotone branches by their filled-in versions which map onto the restrictive interval. The resulting saturated maps are quasiconformally branchwise equivalent provided the terminal box mappings were so. In the infinitely renormalizable case, we obtain infinitely many branchwise equivalent pairs of saturated maps. Our aim is to combine the branchwise equivalences into one quasiconformal map (see Figure 2.2) with uniformly bounded distortion. This is by no means obvious and must be done with special care in order not to accumulate quasiconformal dilatation. Generally, pasting quasiconformal maps increases dilatation by an additive constant. In our setting, however, every dilatation added on different stages are confined to disjoint sets and so do not add up.

The Feigenbaum map is a prototype infinitely renormalizable map and its combinatorics and dynamics are particularly simple. It has zero topological entropy but infinitely many periodic points, all of them of periods 2^n, $n = 1, 2, \ldots$.

In [39], a class of infinitely renormalizable maps with bounded combinatorics was studied and particularly, the Main Theorem was established for the Feigenbaum combinatorics. The key observation was that the forward orbits of restrictive intervals are subjected to bounded combinatorics which yields quasi-symmetric conjugacy on the critical orbits. Our approach here is different and does not rely on the global dynamics. It rather follows Theorem 2 from [24].

In the Feigenbaum case, Theorem 2.1 becomes almost trivial and the Reduced Theorem is an immediate consequence from the polynomial-like property and Proposition 2. This gives a complete proof of the Main Theorem and we plan to use the Feigenbaum case to illustrate underlying concepts of the whole proof. The reader is encouraged to check that our ad hoc proof of Theorem 2.1 in this chapter works well also for infinitely renormalizable maps with bounded combinatorics.

2.1   Extendibility and Distortion

2.1.1   Distortion lemmas

The Schwarzian derivative of a C^3 local diffeomorphism is given by

$$Sf := \frac{f'''}{f'} - \frac{3}{2}(\frac{f''}{f'})^2 \; .$$

There is a simple formula for the Schwarzian of a composition

$$S(f \circ g) = Sf \circ g \cdot (g')^2 + Sg$$

which for iterates of f becomes

$$Sf^n = \sum_{i=0}^{n-1} (Sf \circ f^i) \cdot ((f^i)')^2 \; .$$

Clearly, the Schwarzian of $f_a = a(1 - x^2) - 1$ is negative and the same remains true, by the composition formula, for any monotone branch of a saturated map induced by f_a.

The Real Köbe Lemma.

Definition 2.1.1 *Consider a diffeomorphism h onto its image (b, c). Suppose that its has an extension $\tilde{h}$ onto a larger image (a, d) so that $\tilde{h}$ is still a diffeomorphism. Provided that $\tilde{h}$ has negative Schwarzian derivative, and $\frac{|a-b| \cdot |c-d|}{|c-a| \cdot |d-b|} \geq \epsilon$, we will say that h is ϵ-extendible.*

The following holds for ϵ-extendible maps:

Fact 2.1.1 *There is a function C of ϵ only so that $C(\epsilon) \to 0$ as $\epsilon \to 1$ and for every diffeomorphism h defined on an interval I and ϵ-extendible,*

$$|h''/h'(x)| \cdot |I| \leq C(\epsilon)$$

for every x in the domain of h.

Proof:

Apart from the limit behavior as ϵ goes to 1, this fact is proved in [30], Theorem IV.1.2 . The asymptotic behavior can be obtained from Lemma 1 of [18] which says that if $\tilde{h}$ maps the unit interval into itself, then

$$|h''/h'(x)| \leq \frac{2h'(x)}{\text{dist}\,(\{0,1\}, h(x))} \; . \tag{2.1}$$

The condition $\tilde{h}(0,1) = (0,1)$ can be satisfied by pre- and post-composing $\tilde{h}$ with affine maps. This will not change $|h''/h'(x)| \cdot |I|$, so we just assume that $\tilde{h}$ is normalized in this way. For $\epsilon \geq 1/2$, $|h'(x)|$ is no more than

$$\exp(C(\tfrac{1}{2})) \frac{|h(I)|}{|I|} \; .$$

Hence,

$$|h''/h'(x)| \cdot |I| \leq \exp(C(\tfrac{1}{2})) \frac{|h(I)|}{\text{dist}(h(I), \{0,1\})} \leq \exp(C(\tfrac{1}{2}))(\tfrac{1}{\epsilon} - 1)$$

where the last estimate follows since

$$\frac{\text{dist}(h(I), \{0,1\})}{\text{dist}(h(I), \{0,1\}) + |h(I)|} \geq \epsilon$$

by the extendibility condition.

$\square$

Extendibility of monotone branches. The geometric condition of the extendibility in Definition 2.1.1 amounts to saying that the length of the range (b, c) in the Poincaré metric of the extended range (a, d) is no more than $-\log \epsilon$. [1]

---

[1] See [30] for a discussion of the Poincaré metric on the interval.

2.1.2   Geodesic neighborhoods

This convenient tool was introduced in [39]. Our proofs follow [30] with some modifications.

Consider an interval $[x - y, x + y]$, $y > 0$, on the real line.

Definition 2.1.2 *Look at two circles passing through $x-y$ and $x+y$, one centered at $x + it$, and the other at $x - it$ where $t \in \mathbb{R}$. Suppose that the circles intersect the real line at angle $\alpha \leq \pi/2$. Consider the two open disks delimited by these circles. The union of these disks will be called the* geodesic neighborhood *of $[x - y, x + y]$ with angle $\pi - \alpha$, and denoted $D(\pi - \alpha, [x - y, x + y])$. Likewise, the intersection of these disks will be called the* geodesic neighborhood *of $[x-y, x+y]$ with angle α and denoted $D(\alpha, [x - y, x + y])$.*

The following properties make the geodesic neighborhoods convenient.

Fact 2.1.2 *Let h be a an analytic diffeomorphism with range U and domain V, both intervals on the real line. Assume further that h^{-1} has an analytic continuation as a univalent mapping from the upper half-plane into itself. Then, for any $\alpha \in (0, \pi)$ the preimage of $D(\alpha, U)$ by the analytic continuation of h is contained in $D(\alpha, V)$.*

Proof:
Given an interval W on the real line, consider the set J_W defined as

$$\bigcup_{0 < \alpha < \pi} D(\alpha, W) .$$

The region J_W has its hyperbolic metric ρ_W. Our basic observation is that for any α there is a unique k so that the geodesic neighborhood with angle α can be characterized as follows:

$$D(\alpha, W) = \{z \in J_W : \rho_W(z, W) < k\} .$$

It is clear that this property does not depend on W, so let us assume that $W = [-1, 1]$. The transformation

$$\theta(z) = \frac{z + 1}{z - 1}$$

maps the set

$$\hat{\mathbb{C}} \setminus \{x + i \cdot 0 : x \geq 0\}$$

onto J_W. Thus, the mapping $z \to \theta(z^2)$ provides the uniformizing map from the upper half-plane onto J_W. The preimage of $D(\alpha, W)$ by θ is a wedge-shaped set

$$\{x + iy : -x > \sqrt{x^2 + y^2} \cos \alpha\}$$

and the preimage by the uniformizing map is

$$\{x + iy : y > 0, |x| < y \tan \frac{\alpha}{2}\} .$$

We notice that all points from the set

$$\{x + iy : y > 0, |x| = y \tan \frac{\alpha}{2}\}$$

are in the same Poincarè distance from the imaginary half-line. Indeed, if we take a point in this set, we can map it to any other point by transformations in the form $z \to \lambda z$ with $\lambda > 0$ and the mirror reflection by the imaginary axis. All these transformations fix the imaginary half-line and are hyperbolic isometries, so our assertion follows.

To conclude the proof, observe that h^{-1} is defined as an analytic mapping on the entire J_U. This inverse mapping is into J_V, thus it does not expand the hyperbolic metric. The claim now follows from our characterization of geodesic neighborhoods.

$\square$

Fact 2.1.3 *Let h be a an analytic diffeomorphism with range U and domain V, both intervals on the real line. Assume further that h^{-1} has an analytic continuation as a univalent mapping from the upper half-plane into itself. Suppose also that h is still a diffeomorphism from a larger interval $I' \supset I$ onto a larger interval $J' \supset J$ so that the Poincaré length of J inside J' does not exceed $- \log \epsilon$. Then, for every $\alpha < \pi$ and ϵ there is constant K_1 so that for every pair of points $z_1, z_2 \in D(\alpha, J)$*

$$|\log \frac{|z_1 - z_2|}{|h^{-1}(z_1) - h^{-1}(z_2)|} - \log \frac{|J|}{|I|}| < K_1 .$$

Proof:
The inverse branch of h which maps J' onto I' is well defined in the upper, as well as lower, half-plane. Now, the geodesic neighborhood of J with angle α is contained in the union of all geodesic neighborhoods of J' with modulus bounded away from 0 in terms of ϵ and α. The claim follows by Köbe's distortion lemma.

□

2.2   Saturated Maps

Construction of saturated maps.   We will develop the technique of pasting sequences of branchwise equivalences, finite or not, together in order to get a quasiconformal conjugacy between the postcritical orbits of the corresponding quadratic polynomials.

Consider a terminal type I box mapping obtained from f in the way described above. The forward critical orbit of f is confined to a finite cycle of intervals, i.e. the images of the restrictive interval. These intervals can be contained in preimages of the restrictive interval. Each of these preimages is mapped onto the restrictive interval diffeomorphically. Moreover, it is known that the distortion of this diffeomorphism is bounded by a constant. We can express these properties in terms of *saturated maps* as follows.

Definition of saturated maps.   Consider a mapping $\varphi : V \to \mathbb{C}$ defined on an open set $V \subset \mathbb{R}$, usually disconnected. As usual a *branch* of φ will mean restriction of φ to a connected component of its domain.

Definition 2.2.1 *A mapping φ defined on some set $V \subset \mathbb{R}$ into $\mathbb{R}$ is called* saturated *with bound $K > 0$ if the following conditions are met:*

- *φ has only monotone branches with one common range U,*

- *for every branch of φ, its inverse has an analytic continuation to a univalent transformation of the upper half-plane into $\mathbb{C} \backslash \mathbb{R}$,*

- *all branches of φ are K^{-1} extendible (see Definition 2.1.1).*

Let us next consider the notion of *nesting* saturated mappings.

Definition 2.2.2 *Let φ be a saturated mapping with domain V and $W \supset V$ an interval. We say that φ is nested in W with bound K if for every connected component D of V*

$$\frac{|D|}{\text{dist}(D, \partial W)} < K .$$

Typically, W will be chosen as a convex hull of the domain of ϕ. In particular, the domain of ϕ will accumulate towards the endpoints of W. The simplicity of the Feigenbaum case, more generally infinitely renormalizable case with bounded combinatorics, is amplified by fact that one can choose W so that the domain of ϕ be compactly contained in W on every stage of the gluing construction.

Equivalence of saturated mappings. We introduce an analogon of the "branchwise equivalence" between saturated maps.

Definition 2.2.3 *Let φ and $\hat{\varphi}$ be saturated maps with ranges U and $\hat{U}$, respectively. Specify intervals $W \supset U$ and $\hat{W} \supset \hat{U}$ and an order-preserving homeomorphism v from U onto $\hat{U}$. We say that an order-preserving homeomorphism h of the real line is a branchwise equivalence between φ and $\hat{\varphi}$ given $(W, \hat{W}, v)$ if the following is satisfied:*

- *h maps W onto $\hat{W}$ and the domain of φ onto the domain of $\hat{\varphi}$,*

- *$v \circ \varphi(x) = \hat{\varphi} \circ h(x)$ for every x in the domain of φ.*

Regularity of saturated pairs. A branchwise equivalence h with parameters $(W, \hat{W}, v)$ is just an extension of a piecewise continuous mapping whose branches are equal to $\hat{\varphi}^{-1} \circ v \circ \varphi$ to a homeomorphism mapping W onto $\hat{W}$. It cannot, therefore, be expected to be quasi-symmetric with a better bound than v. It also maps the domain of φ to the domain of $\hat{\varphi}$ and this imposes a constraint independent of a choice of v. So it is reasonable to require that for every v, a quasi-symmetric branchwise equivalence given v exists, and its bound depends on the bound of v in some uniform way. Our next definition says this and a little more, namely that outside of W these branchwise equivalences should be independent of v (see Figure 2.2).

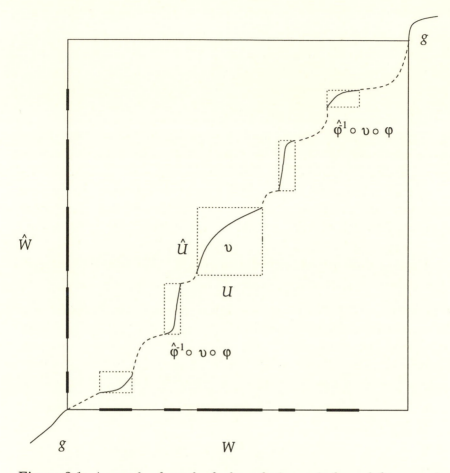

Figure 2.1: A sample of graph of a branchwise equivalence h for a regular pair $(\varphi, \hat{\varphi})$ of saturated maps. In dotted boxes, there are pieces of the graph of $\hat{\varphi}^{-1} \circ \upsilon \circ \varphi$ and outside of the big box, there is the graph of g. Be aware that typically there are infinitely many dotted boxes with pieces of the graph of $\hat{\varphi}^{-1} \circ \upsilon \circ \varphi$. The domains of $\hat{\varphi}$ and φ are drawn in bold line.

Definition 2.2.4 *Given a pair $(\varphi, \hat{\varphi})$ of saturated mappings choose intervals W and $\hat{W}$ so that φ and $\hat{\varphi}$ are nested with bound K in $\hat{W}$ and W, respectively. Let $L : [1, \infty) \to \mathbb{R}$ be a function. We say that φ and $\hat{\varphi}$, given W and $\hat{W}$, form an $L(K)$-regular pair if there is an order-preserving homeomorphism g from $\mathbb{R} \backslash W$ onto $\mathbb{R} \backslash \hat{W}$ so that for every K-quasi-symmetric order-preserving homeomorphism υ of the range of φ onto the range of $\hat{\varphi}$ there is an $L(K)$-quasi-symmetric* *branchwise equivalence h between φ and $\hat{\varphi}$ given $(W, \hat{W}, v)$ which equals g outside of W.*

Regular pairs of saturated maps.

Theorem 2.1 *Let f and $\hat{f}$ be unimodal quadratic polynomials, topologically conjugate and critically recurrent. If f and $\hat{f}$ are non-renormalizable then the claim of the Reduced Theorem holds. Otherwise, there are saturated mappings φ and $\hat{\varphi}$ induced by f and $\hat{f}$, respectively, with the following properties.*

- *The domain of φ contains the forward critical orbit of f and the domain of $\hat{\varphi}$ contains the forward critical orbit of $\hat{f}$.*

- *The image of all branches of φ is the maximal restrictive interval of f, and the image of all branches of $\hat{\varphi}$ is the corresponding maximal restrictive interval of $\hat{f}$.*

- *Both φ and $\hat{\varphi}$ are saturated and nested in $(-1, 1)$ with bound K_1 (see Definitions 2.2.2 and 2.1.1) where K_1 is a constant independent from f and $\hat{f}$.*

- *The pair $(\phi, \hat{\phi})$ is $L(K)$-regular given $(-1, 1)$ and $(-1, 1)$ where L is a fixed function, independent of f and $\hat{f}$.*

The main claim of Theorem 2.1 is the last one. It will be derived relatively easily from the fact that there is a quasiconformal branchwise equivalence with maximal dilatation bounded by a constant between the terminal maps. Theorem 2.1 was proved before for S-unimodal mappings with a particular "basic-renormalizable" dynamics in [24]. As was mentioned before the Misiurewicz case is not covered by the Reduced Theorem. The following lemma is proved implicitly in Chapter 5. We encourage the reader to supply details in this simple case.

Fact 2.2.1 *Let f and $\hat{f}$ be unimodal quadratic polynomials, topologically conjugate. Assume that all periodic points are repelling and the critical orbits are not recurrent. Then the claim of the Reduced Theorem also holds.*

The full proof of Theorem 2.1 will be given in Chapter 5.

Proof of Theorem 2.1 in the Feigenbaum case.   In the proof
of Theorem 2.1, Chapter 5, we distinguish two cases in dependence
on whether f has an odd orbit or not. The Feigenbaum map clearly
falls into a second category.

Let f and $\hat{f}$ be unimodal quadratic polynomials with Feigenbaum
dynamics. The first return map onto the fundamental inducing domain D_0 of f has only one branch. We define a saturated map φ by
the formula

$$\varphi(x) = \begin{cases} x & \text{if } x \in D_0 \\ f_r & \text{if } x \in D_{-1} = f_r^{-1}(D_0) \end{cases},$$

where f_r is the right lap of f. The saturated map $\hat{\varphi}$ is defined analogously. It is well known that for the Feigenbaum dynamics the
lengths of D_0 and D_{-1} are bounded away from 0 and 2 independently from f and the absolute value of the derivative of f_r on D_{-1}
is bounded away from zero and infinity by a constant independent
from f (we have short proofs of these statements for quadratic polynomials with infinitely many periodic points based on λ-Lemma in
Chapter 5).

$L(K)$-regularity property.   The $L(K)$-regularity given the intervals $(-1,1), (-1,1)$ is not hard to check directly. Let v be a K-
quasi-symmetric order-preserving homeomorphism between restrictive intervals D_0 and $\hat{D}_0$. The homeomorphism h is defined by the
formula

$$h(x) = \begin{cases} v(x) & \text{if } x \in D_0 \\ \hat{f}_r^{-1} \circ v \circ f(x) & \text{if } x \in D_{-1} \end{cases}.$$

Clearly, h restricted to D_{-1} is quasi-symmetric with the bound
depending on K and the distortion of f and $\hat{f}$. This is, however,
not enough to guarantee that h is quasi-symmetric on $D_0 \cup D_1$ since
points are not removable for quasi-symmetric mappings.

We proceed further to see that by bounded distortion,

$$h(q + x) = k'(x)h(q - k(x)x)$$

for every $x > 0$ and $k(x), k'(x)$ are bounded from above and below by
positive constants independent of f and $\hat{f}$. It follows that h is quasi-
symmetric with the bound independent of f and $\hat{f}$. By Lemma 3.12

of [24], h has an extension to the real line which is identity outside of $(-1, 1)$ and is quasi-symmetric with the similarly uniform bound.

Generalization of Theorem 2.1. It follows that the claim of Theorem 2.1 remains valid in the following situation.

Let F and $\hat{F}$ be topologically conjugate, critically recurrent and renormalizable polynomials. Choose a locally maximal restrictive interval I for F, and the corresponding restrictive interval $\hat{I}$ for $\hat{F}$. Then define f to be the first return map of F on I and $\hat{f}$ is defined analogously.

Then the claim of Theorem 2.1 remains true, Moreover, the bounds K_1 and $L(K)$ are independent of the choice of I.

To prove this, observe that by Theorem 1.1, f is conjugate to a unimodal polynomial by a quasi-symmetric map H with a constant bound. The same is true of $\hat{f}$ with a conjugating homeomorphism $\hat{H}$. Let P and $\hat{P}$ denote these polynomials. Let φ_P and $\hat{\varphi}_P$ be the saturated map constructed by Theorem 2.1 when applied to P and $\hat{P}$. Then

$$\varphi := H^{-1} \circ \varphi_P \circ H$$

is a saturated map for f and

$$\hat{\varphi} := \hat{H}^{-1} \circ \hat{\varphi}_P \circ \hat{H}$$

is a saturated map for $\hat{f}$. The first two "topological" claims of Theorem 2.1 are obvious. The last claim is also immediate, with the bound $L(K)$ modified depending on the quasi-symmetric estimates for H and $\hat{H}$. The bound remains independent of f, $\hat{f}$ or the choice of I. The third claim is also valid. The nesting condition of Definition 2.2.2 is quasi-preserved by conjugacies H and $\hat{H}$, again with uniform bounds. So is the extendibility bound of Definition 2.2.1.

2.3   Gluing of Saturated Maps

Chains of saturated mappings. Observe that if two saturated mappings φ_1 and φ_2 are given and the range of φ_1 contains the domain of φ_2, then $\varphi_2 \circ \varphi_1$ is another saturated map.

Proposition 2 *Let φ_1, .., φ_n and $\hat{\varphi}_1$, .., $\hat{\varphi}_n$ be sequences of saturated mappings with ranges U_i and $\hat{U}_i$ respectively. Specify two intervals U_0 and $\hat{U}_0$. For every i between 1 and n we assume that*

- *φ_i and $\hat{\varphi}_i$ are saturated and nested in U_{i-1} and $\hat{U}_{i-1}$, respectively, with bound K_1,*

- *the pair $(\varphi_i, \hat{\varphi}_i)$ is $L(K)$-regular given U_{i-1} and $\hat{U}_{i-1}$.*

Then for every K_1 and every function $L(K)$ there is a function $L'(K)$ so that the pair

$$(\varphi_n \circ \cdots \circ \varphi_1, \hat{\varphi}_n \circ \cdots \circ \hat{\varphi}_1)$$

is $L'(K)$-regular given $U_0, \hat{U}_0$.

Proposition 2 is very close to Theorem 2 of [24] in the complex analytic rather than S-unimodal setting. Figure 2.2 shows schematically how to glue branchwise equivalences between two sequences $\{\phi_i\}_{i=1}^n$ and $\{\hat{\phi}_i\}_{i=1}^n$ of the corresponding saturated mappings.

Preliminaries. In this section we will invoke a few times Theorem 2.1 with f and $\hat{f}$ equal to the renormalizations of f_i and $\hat{f}_i$. This means that we really use the generalization of Theorem 2.1 introduced in Section 2.2. "Uniform constants" in the proof will be numbers dependent on the parameter K_1 and the function $L(K)$ from Proposition 2.

We will use the following analytic fact.

Fact 2.3.1 *Let h be an L-quasi-symmetric orientation-preserving homeomorphism from the real line onto itself mapping an interval U onto its image $\hat{U}$. For every L, there is L_1 and an L_1-quasiconformal homeomorphism H from the complex plane onto itself which commutes with complex conjugation, i.e. $\overline{H(z)} = H(\bar{z})$, extends h, and maps $\mathcal{D}(U, \frac{\pi}{4})$ onto $\mathcal{D}(\hat{U}, \frac{\pi}{4})$.*

Proof:
The proof can be done "by hand" with the use of Beurling and Ahlfors' extension theorem. We leave it out as an exercise.

$\square$

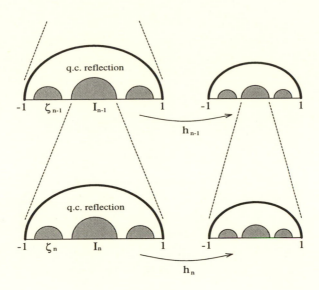

Figure 2.2: Construction of gluing branchwise equivalences between the corresponding sequences of saturated maps.

In general, branchwise equivalences do not have to preserve the real line. The following lemma shows that it is always possible to modify them to satisfy this.

Lemma 2.3.1 *Let D and $\hat{D}$ be Jordan domains symmetric with respect to the real line and bounded by Jordan curves. Let H be a Q-quasiconformal homeomorphism of the complex plane onto itself which maps D onto $\hat{D}$ and satisfies $\Upsilon(\bar{z}) = \overline{\Upsilon(z)}$ for z on the boundary of D. Then, there is a $2Q$-quasiconformal homeomorphism Υ' of the plane which coincides with Υ outside D and maps $D \cap \mathbb{R}$ onto $\hat{D} \cap \mathbb{R}$.*

Proof:
Let h be a Riemann mapping from the upper-half plane onto D, symmetric with respect to the imaginary line and sending 0 and ∞ to the points where the boundary of D intersects the real line. The Riemann mapping $\hat{h}$ is chosen in the analogous way for $\hat{D}$. Since

both Riemann maps extend to the boundary, then

$$\Upsilon_1 := \tilde{h}^{-1} \circ \Upsilon \circ h$$

is continuous in the closure of the upper half-plane. In the first quarter of the complex plane consider the mapping R_1 given by

$$(r, \theta) \to (r, 2\theta)$$

in the polar coordinates. In the second quarter, consider R_2 given by

$$(r, \theta) \to (r, 2\theta - \pi) \, .$$

Both R_1 and R_2 are 2-quasiconformal and map their respective domains onto the upper half-plane. If we consider mappings $R_2^{-1} \circ \Upsilon_1 \circ R_2$ and $R_1^{-1} \circ \Upsilon_1 \circ R_1$, then we see that they match along the imaginary axis as the consequence of Υ_1 being odd on the real line. Hence, they can be glued together to a $2Q$-quasiconformal map Υ_2 which coincides with Υ_1 on the real line. Then we make Υ' equal to $\hat{h} \circ \Upsilon_2 \circ h^{-1}$ on D.

$$\square$$

2.3.1   The main step of the construction

The following lemma describes the main construction leading to the proof of Proposition 2. Let us introduce a geometric concept first.

Definition 2.3.1 *If φ is a saturated map with range U, define its geodesic continuation $T(\varphi)$ by continuing each branch analytically onto $D(U, \frac{\pi}{4})$.*

Note that for each branch such a continuation is possible and univalent by Definition 2.2.1 and if the domain of the branch is S, then the domain of its geodesic continuation is contained in $\mathcal{D}(S, \frac{\pi}{4})$ by Fact 2.1.2.

Dilatation confined to disjoint sets.   Consider a pair of saturated maps φ and $\hat{\varphi}$ with ranges U and $\hat{U}$, respectively. Choose intervals W and $\hat{W}$ containing the domains of φ and $\hat{\varphi}$, respectively, and so that the pair $(\varphi, \hat{\varphi})$ is $L(K)$-regular given W and $\hat{W}$. We claim that:

Lemma 2.3.2 *There exist a constant M_0, which depends only on $L(1)$, and an M_0-quasiconformal homeomorphism Υ_0 of $\mathbb{C}$ which is symmetric about $\mathbb{R}$ and maps*

$$\mathbb{C} \setminus \mathcal{D}(W, \frac{\pi}{4}) \xrightarrow{\text{onto}} \mathbb{C} \setminus \mathcal{D}(\hat{W}, \frac{\pi}{4}),$$

and satisfies the following gluing property : Suppose that an M^∞-quasiconformal homeomorphism of

$$\mathbb{C} \setminus \mathcal{D}(U, \frac{\pi}{4}) \xrightarrow{\text{onto}} \mathbb{C} \setminus \mathcal{D}(\hat{U}, \frac{\pi}{4})$$

extends to an M_{-1} quasiconformal homeomorphism Υ_{-1} of the plane, symmetric with respect to the real line.

Then there exist a constant M, which depends only on M^∞ and the gauge function $L(K)$, and a $\max(M_{-1}, M)$-quasiconformal homeomorphism Υ of $\mathbb{C}$, symmetric about $\mathbb{R}$, and such that

- *Υ coincides with Υ_0 outside $\mathcal{D}(W, \frac{\pi}{4})$,*

- *$T(\hat{\varphi}) \circ \Upsilon = \Upsilon_{-1} \circ T(\varphi)$,*

where T stands for the geodesic continuation (see Definition 2.3.1).

Proof:
Recall Definition 2.2.4 of regular pairs and apply to φ and $\hat{\varphi}$ with v affine. This yields an $L(1)$-quasi-symmetric branchwise equivalence between them equal to the fixed g outside of W. This branchwise equivalence can be continued to Υ_0 using Fact 2.3.1.

The map Υ is then already determined on the domain of the geodesic continuation, that is on

$$\mathcal{T} = \bigcup_\zeta \zeta^{-1}(\mathcal{D}(U, \frac{\pi}{4}))$$

where ζ ranges over the set of all branches of φ, as well as on the complement of $\mathcal{D}(W, \frac{\pi}{4})$ where it is equal to Υ_0. The resulting interpolation problem is quite difficult because the set $\mathcal{T}$ need not be relatively compact in $\mathcal{D}(W, \frac{\pi}{4})$. The solution is possible based on our notion of regular pairs of saturated maps.

It is enough to construct Υ and prove the needed estimate on its Beltrami bound in the upper half plane and then extend it by reflection. As the first step, we take Υ_{-1} and change it inside $\mathcal{D}(U, \frac{\pi}{4})$

to make it K_1-quasiconformal where K_1 only depends on M^∞ and not M_{-1}. This is done easily by reflecting Υ_{-1} restricted to the complement of $\mathcal{D}(U, \frac{\pi}{4})$ in the boundary of $\mathcal{D}(U, \frac{\pi}{4})$. Let us call the resulting map Υ_1'. By Lemma 2.3.1, without loss of generality we may assume that Υ_{-1}' is symmetric about the real line. Let v_{-1}' denote the $K_2(M^\infty)$-quasi-symmetric homeomorphism of U onto $\hat{U}$ defined by restricting Υ_{-1}' to the real line. The condition of regularity (Definition 2.2.4) applied to φ and $\hat{\varphi}$ with $v := v_{-1}'$ yields a K_3-quasi-symmetric homeomorphism h of the whole line which is a branchwise equivalence given $(W, \hat{W}, v_{-1}')$ and is equal to the fixed g outside of W. In other words, h coincides with what Υ has to be on $\mathcal{T} \cap \mathbb{R}$ and $\mathbb{R} \setminus W$. We construct the homeomorphism H as follows. Let H

- satisfy $T(\hat{\varphi}) \circ H = \Upsilon_{-1}' \circ T(\varphi)$ on $\mathcal{T}$ intersected with the upper half-plane $\mathbb{H}^+$,

- be a Beurling-Ahlfors extension of h in the lower half plane $\mathbb{H}^-$,

- equal Υ_0 in the complement of $\mathcal{D}(W, \frac{\pi}{4})$ in $\mathbb{H}^+$.

By construction, H is continuous on its domain and K_4-quasiconformal in its interior, where K_4 depends on M^∞ and $L(K)$ but not M_{-1} and extends continuously to the border of $\mathcal{T}$ contained in $\mathbb{H}^+$ in the same way as Υ defined on $\mathcal{T}$ by the condition

$$T(\hat{\varphi}) \circ \Upsilon = \Upsilon_{-1} \circ T(\varphi) .$$

The last observation is true since Υ_{-1}' and Υ_{-1} are equal on the boundary of $\mathcal{D}(U, \frac{\pi}{4})$. Observe that, H and Υ are generally not equal on $\mathcal{T}$ since H is the pull-back of Υ_{-1}'.

Our next goal is to continue H quasiconformally to the whole plane. This includes defining it on the region Γ equal

$$\mathcal{D}(W, \frac{\pi}{4}) \setminus \mathcal{T}$$

intersected with the upper half-plane. An analogous object, $\hat{\Gamma}$, can be defined in the phase space of $\hat{\varphi}$. We will prove in a moment that both $\partial \Gamma$ and $\partial \hat{\Gamma}$ are quasi-circles with a dilatation bounded by a constant M' independent of M_{-1} or M^∞. Right now, just assume this. Then H can be extended to the whole plane by a quasiconformal reflection in Γ and $\hat{\Gamma}$. Then H is M-quasiconformal in the plane where M depends only on M^∞ and the regularity gauge $L(K)$.

This H is still not equal to Υ given by

$$T(\hat{\varphi}) \circ \Upsilon = \Upsilon_{-1} \circ T(\varphi)$$

on $\mathcal{T} \cap \mathbb{H}^+$. But Υ and H coincide on the boundary of each connected component of $\mathcal{T} \cap \mathbb{H}^+$, so one can just replace H by Υ in these regions to get the needed extension of Υ in the upper-half plane. This proves Lemma 2.3.2 under the extra hypotheses about the quasiconformal of Γ and $\hat{\Gamma}$.

Quasiconformality of arcs. We quote a standard definition of a quasiconformal curve following [25], page 97:

Definition 2.3.2 *A Jordan curve w in the Riemann sphere is called K-quasiconformal if there is a K-quasiconformal homeomorphism H of $\hat{\mathbb{C}}$ onto itself which maps w onto the unit circle.*

A very convenient criterion for checking whether a Jordan curve is K-quasiconformal, was given by Ahlfors, see [25] where it is called "bounded turning". We follow a version from [11] which mentions estimates between the constant appearing in the bounded turning property and the K. The book [11] uses the name "three-point property" rather than "bounded turning".

Fact 2.3.2 *Let w be a Jordan curve in the Riemann sphere. For every pair of points z_1 and z_2 consider u_1 and u_2 defined as diameters, in the spherical metric, of the two arcs cut from w by z_1 and z_2. Define*

$$Z := \sup\left\{ \frac{\min(u_1, u_2)}{\mathrm{dist}(z_1, z_2)} : z_1, z_2 \in w \right\}.$$

For every finite Z there is a constant Q so that w is a Q-quasiconformal curve. Conversely, if w is Q-quasiconformal, then Z is bounded in terms of Q.

The first statement saying that K is bounded in terms of Z follows from Theorem 6.2, page 63, in [11]. The converse statement requires the reader to follow a longer route through Theorems 5.1 on page 57, 4.1 from page 55 and 2.3 on page 51, ibidem.

Conclusion of the proof. We will only prove that $\partial\Gamma$ is quasiconformal with dilatation bounded by M' since the proof for $\partial\hat\Gamma$ is quite the same. Inside the boundary of Γ, distinguish $\partial\Gamma^+$ which is the boundary of $\mathcal{D}(U, \frac{\pi}{4})$ intersected with $\mathbb{H}^+$ and $\partial\Gamma^- = \partial\Gamma \setminus \partial\Gamma^+$ which is the union of disjoint analytic arcs, and segments of the real line which join these arcs. These real segments can be also characterized as the complement of the domain of φ in W.

The main tool is the bounded turning property, stated here as Fact 2.3.2. Since we are working in a bounded region, the spherical metric is equivalent to the Euclidean one. In the setting of Fact 2.3.2, we pick two points z_1 and z_2 on $\partial\Gamma$ and we need to bound the quantity Z. If both z_1 and z_2 are on $\partial\Gamma^+$, the bound is obvious.

Suppose that z_1 and z_2 are both on $\partial\Gamma^-$. The first possibility is that they are both in the boundary of the same connected component of $\mathcal{T}$. But this arc is the preimage of the border of $\mathcal{D}(U, \frac{\pi}{4})$ by the analytic continuation of a branch ζ. The distortion of ζ is bounded in terms of its extendibility, thus by a uniform constant. So the quantity Z remains bounded by a uniform constant. If z_1 and z_2 are in the boundaries of different connected components, or on the real line, then we use the fact that each such component is contained in the geodesic neighborhood of angle $\pi/4$ of its intersection with the real line. Therefore, the diameter of the arc u_1 joining z_1 and z_2 inside $\partial\Gamma^-$ is bounded by $\sqrt{2}$ times the distance between the projections of z_1 and z_2 on the real line, hence also by $\sqrt{2}|z_1 - z_2|$.

If z_1 is on $\partial\Gamma^+$ and z_2 is on $\partial\Gamma^-$, and the arc u_1 joining them passes through an endpoint a of W, then

$$\frac{\operatorname{diam}(u_1)}{\operatorname{dist}(z_1, a) + \operatorname{dist}(z_2, a)} \le M_1 . \qquad (2.2)$$

This is because $u_1 \cap \partial\Gamma^-$ and $u_1 \cap \partial\Gamma^+$ are both quasiconformal with dilatation bounded by a uniform constant, as shown above. Next, we use the information that $\partial\Gamma^+$ and $\partial\Gamma^-$ are separated by the boundary of a geodesic neighborhood of W with angle $\epsilon < \frac{\pi}{4}$ where

ϵ is a uniform constant. Indeed, ϵ is given in terms of the nesting bound K_1 from the hypothesis of Proposition 2. This implies that

$$\text{dist}(z_2, \Gamma^+) \geq M_2(\epsilon)\text{dist}(z_2, a)$$

and

$$\text{dist}(z_1, \Gamma^-) \geq M_2(\epsilon)\text{dist}(z_1, a)$$

provided that z_1 and z_2 are closer to a then to the other endpoint of W and with $M_2(\epsilon)$ a uniform constant. So

$$\text{dist}(z_1, z_2) \geq \frac{1}{2}M_2(\epsilon)(\text{dist}(z_1, a) + \text{dist}(z_2, a)) \ .$$

In view of estimate (2.2) this leads to

$$\frac{\text{diam}\,(u_1)}{\text{dist}(z_1, z_2)} \leq \frac{2M_1}{M_2(\epsilon)} \ .$$

Hence if the projections of z_1 and z_2 are in the same half of W, we are done. Otherwise, the distance from z_1 to z_2 is bounded from below in comparison with $|W|$.

$\square$

The inductive argument. As the first step, use Lemma 2.3.2 with the gauge function $L(K)$ specified in Proposition 2 to obtain the constant M_0. Then choose m by introducing $M^\infty := M_0$, the same $L(K)$, and setting $m = M$. The constant m depends only on $L(K)$.

Suppose now that we are in the situation of Proposition 2. To verify the regularity condition for the compositions of saturated maps, begin with a K-quasi-symmetric homeomorphism v from U_n onto $\hat{U}_n$. By Fact 2.3.1, this v can be extended to an quasiconformal homeomorphism Υ_{-1} of the plane, symmetric about the real line and mapping $\mathcal{D}(U_n, \frac{\pi}{4})$ onto $\mathcal{D}(\hat{U}_n, \frac{\pi}{4})$. Then use Lemma 2.3.2 with $\varphi := \varphi_n$, $\hat{\varphi} := \hat{\varphi}_n$ and Υ_{-1} as specified. This gives a map Υ_{n-1} which is globally $M(K)$-quasiconformal and M_0-quasiconformal as a mapping of $\mathbb{C} \setminus \mathcal{D}(U_{n-1}, \frac{\pi}{4})$ onto itself. This Υ_{n-1} restricted to the real segment U_{n-1} is a branchwise equivalence between φ_n and $\hat{\varphi}_n$ given U_{n-1}, $\hat{U}_{n-1}$ and v.

Hence, by induction, Lemma 2.3.2 immediately implies the following claim, which for $k = n$ yields Proposition 2.

Inductive claim. *For every $1 \leq k \leq n$ there is a $\max(M(K), m)$ quasiconformal homeomorphism Υ_{n-k} of the complex plane, symmetric about the real line, and mapping*

$$\Upsilon_{n-k} : \mathcal{D}(U_{n-k}, \frac{\pi}{4}) \xrightarrow{\text{onto}} \mathcal{D}(\hat{U}_{n-k}, \frac{\pi}{4}) \ .$$

In addition, Υ_{n-k} is M_0-quasiconformal on $\mathbb{C} \setminus \mathcal{D}(U_{n-k}, \frac{\pi}{4}))$ and the restriction of Υ_{n-k} to the real line is a branchwise equivalence between $\varphi_n \circ \cdots \circ \varphi_{n-k+1}$ and $\hat{\varphi}_n \circ \cdots \circ \hat{\varphi}_{n-k+1}$ given $(U_{n-k}, \hat{U}_{n-k}, \upsilon)$.

2.3.2   Proof of the reduced theorem

If f and $\hat{f}$ are conjugate renormalizable polynomials, let $I_1 \supset \cdots$ be the sequence, finite or not, of all its locally maximal restrictive intervals. Let I_0 denote $(-1, 1)$. Then f_i and $\hat{f}_i$ denote the first return maps into I_i and $\hat{I}_i$, respectively. For every $i \geq 0$, save if I_i is the last in the sequence, we get a pair of saturated mappings described by the generalization of Theorem 2.1. This gives us a "chain" of saturated mappings φ_i and $\hat{\varphi}_i$. The range of φ_i is a restrictive interval which also contains the domain of φ_{i+1}. This sequence is infinite if and only if f is infinitely renormalizable.

We can apply Proposition 2 to our chains of saturated mappings induced by f_i. Here we distinguish between the finitely renormalizable situation when this sequence is finite, and infinitely renormalizable. In the finitely renormalizable situation, let us say that f_{n+1} is non-renormalizable. As already explained, we can get a uniformly quasi-symmetric conjugacy between f_{n+1} and $\hat{f}_{n+1}$ and plug it into Proposition 2 as υ. In that case Proposition 2 will give a mapping which is a conjugacy on the domain of the composition $\varphi_n \circ \cdots \circ \varphi_1$, in particular on the forward critical orbit. In the infinitely renormalizable situation, the same result can be arrived at by a limiting process in which we take longer and longer chains of saturated mappings and apply Proposition 2. The mapping υ is then irrelevant and can be made affine. We get a sequence of uniformly quasi-symmetric maps, so we can pick a quasi-symmetric limit. This limit must conjugate the forward critical orbits, since the domains of $\varphi_n \circ \cdots \circ \varphi_1$ shrink uniformly as we take longer and longer chains. This proves the Reduced Theorem.

Chapter 3

Polynomial-Like Property

Let us recall Theorem 1.1. This chapter is devoted to its proof.

Theorem 1.1 *Let f be a renormalizable quadratic polynomial without attracting or indifferent periodic orbits and let $I_1 \supset \cdots$ be the sequence, finite or not, of its locally maximal restrictive intervals (see Definition 1.3.4). Let f_i denote the first return map into I_i. Then, for every i, function f_i is conjugate to a unimodal quadratic polynomial by an L-quasi-symmetric homeomorphism sending I_i to $(-1, 1)$. The constant L is independent of i.*

    Remark: In reality, L is independent of f as well, but we don't need this fact.

3.1   Domains in the Complex Plane

The complex bounds theorem.

Theorem 3.1 *Let f be a a unimodal quadratic polynomial. Let I be a locally maximal restrictive interval of f, with return time $n > 1$. Assume that $|(f^n)'_{|\partial I}| \geq \beta > 1$. Also, for every m, if m is the return time of some restrictive interval of f, perhaps different from I, suppose that f has no neutral or attracting orbit of period less or equal to $4m$.*

    There are Jordan domains U and V, symmetric with respect to the real line, so that $I \subset U \subset \overline{U} \subset V$, f^n is proper of degree 2 from

45

U onto V, and for every $\beta > 1$ there is a constant $K > 0$ so that

$$\operatorname{mod}(V \setminus \overline{U}) \geq K \,.$$

It is emphasized that K is independent of n or f.

Derivation of Theorem 1.1. Before proving Theorem 3.1, let us see how it implies Theorem 1.1. Let us adopt the notations and hypotheses of Theorem 1.1. Observe that if I is one of the sequence I_i, then the parameter β of Theorem 3.1 can be chosen independent of i. This is because of the following:

Fact 3.1.1 *If $f : J \to J$, where J is an interval, is C^2 and its critical points are non-flat, then there exist $\lambda > 1$ and $n_0 \in \mathbb{N}$ so that*

$$|(f^n)'(p)| > \lambda$$

for every periodic point p of f with period $n \geq n_0$.

Proof:
This appears as Theorem B on page 268 in [30]. Let us only recall that a critical point c is called non-flat there meaning that there is a C^2 diffeomorphism ϕ from a neighborhood on 0 onto a neighborhood of c so that $f \circ \phi$ is a monomial.

$\square$

Now Theorem 3.1 tells us that for every i there are Jordan domains U_i and V_i so that

$$I_i \subset U_i \subset \overline{U}_i \subset V_i$$

and f^{n_i} is proper of degree 2 from U_i onto V_i. Importantly, $\operatorname{mod}(V_i \setminus \overline{U}_i) \geq K$ where K is a constant. Theorem 1.1 follows from the *straightening theorem* of Douady-Hubbard. Let us state the theorem.

Fact 3.1.2 *Let f be a quadratic-like mapping, $f : U \to V$, with $\operatorname{mod}(V \setminus \overline{U}) \geq \kappa$ so that the critical value of f and its image are in U. Suppose that $f(\overline{z}) = \overline{f(z)}$, which is also meant to imply that the domain U of f is symmetric with respect to the real line. For every $K > 0$ there is Q so that if $\kappa \geq K$, then there is a Q-quasiconformal homeomorphism H, defined on an open set W, and the following hold:*

- $H(W) = V$,

- $H(\overline{z}) = \overline{H(z)}$ *for every* $z \in W$,

- $F = H^{-1} \circ f \circ H$ *is a real quadratic polynomial.*

The straightening theorem was introduced in [8]. The version we use will be proved in Chapter 5.

About the proof of Theorem 3.1.   Theorem 3.1 was originally proved in [26] and [15]. We follow a more recent argument of Sands in [36]. It is considerably simpler than the previous ones and avoids geometric estimates in the complex plane. There are two features of this proof which make it different from the rest of the book. First, the formalism of *cutting times* is used to describe the combinatorics of the map. This is different from the inducing process used in other chapters. Secondly, the estimates are obtained near the critical value of the map, instead of the critical point.

3.2   Cutting Times

Definition.   In all of this section, let f be a unimodal map. Fix the critical point at 0 and let c denote $f(0)$. We will study the order of some dynamically defined points near c. A prototype geometric setting of the section in the renormalizable case is illustrated on Figure 3.1.

Definition 3.2.1 *An integer $n > 0$ is called a* cutting time *for f if for some $\alpha < c$ the iterate f^n is monotone on (α, c) and $f^n(\alpha) = c$.*

Obviously, only one α with this property may exist for a given n. Hence, α_n is well-defined when n is a cutting time. We will say that a cutting time n is *proper* if all orbits of f with periods not exceeding n are repelling.

Let us see first of all that

Lemma 3.2.1 *If f has a restrictive interval with return time equal to n and all orbits with periods not exceeding n are repelling, then n is a proper cutting time.*

Proof:
Let I denote a locally maximal restrictive interval with return time n. Since there is no attracting orbit of period n, I contains a preimage of 0 by f_n. Then take a in $f(I)$ so that $f^{n-1}(a) = 0$ and see that the condition of Definition 3.2.1 holds with $\alpha := a$.

$\square$

Properties of cutting times.

Lemma 3.2.2 *If $\overline{n} < n$ and both are proper cutting times, then $\alpha_{\overline{n}} < \alpha_n$.*

Proof:

Note that $f^{\overline{n}}$, thus f^n, have a turning point at $\alpha_{\overline{n}}$. So, in order for f^n to be monotone, $\alpha_n \geq \alpha_{\overline{n}}$. There is no equality either, since that would make 0 periodic with period $n - \overline{n}$.

$\square$

If n is any positive integer, let n' denote the largest cutting time smaller than n. Since 1 is always a cutting time, this is well-defined for $n > 1$.

Lemma 3.2.3 *If n is a cutting time greater than 1, then $f^{n-n'-1}$ is monotone on $(f^{n'}(c), c)$.*

Proof:
Observe that $(f^{n'}(c), c)$ is a monotone (with the reversal of order) image of $(\alpha_{n'}, c)$ by $f^{n'}$. Let k be the smallest so that f^k is not monotone on $(f^{n'}(c), c)$. We choose a on $(\alpha_{n'}, c)$ so that $f^{n'+k}(a) = c$. Since there is no cutting time between n' and n, $n - n' - 1 \leq k - 1$ and $f^{n-n'-1}$ is monotone on $(f^{n'}(c), c)$.

$\square$

Lemma 3.2.4 *If n is a proper cutting time greater than 1, then $n - n'$ is a cutting time and $f^{n'}(\alpha_n) = \alpha_{n-n'}$.*

Proof:
Denote $a := f^{n'}(\alpha_n)$. By Lemma 3.2.2, $a \in (f^{n'}(c), c)$. Also $f^{n-n'}(a)$ equals to $f^n(\alpha_n) = c$. From Lemma 3.2.3, $f^{n-n'}$ is monotone on (a, c). Thus, $n - n'$ is a cutting time and $a = \alpha_{n-n'}$.

$\square$

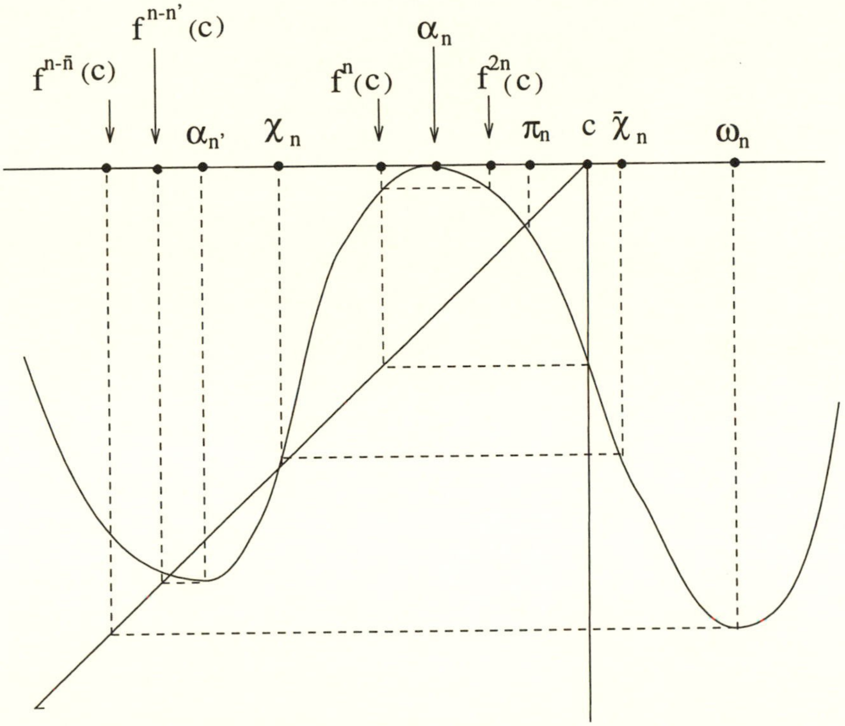

Figure 3.1: Geometry of some dynamical points near the critical value in the renormalizable case.

Lemma 3.2.5 *If $n, \bar{n}$ are proper cutting times and $n > \bar{n}$, then $f^{\bar{n}}(c) < \alpha_{\bar{n}}$.*

Proof:
Suppose that the claim of the lemma does not hold. Then the interval $(\alpha_{\bar{n}}, c)$ is mapped monotonically into itself by $f^{\bar{n}}$. It follows that all iterates of f are monotone on this interval. However, by

Lemma 3.2.2, it contains α_n which is a turning point for f^n, contradiction.

$\square$

Lemma 3.2.6 *If n is a cutting time and f^{2n} has only repelling orbits, then $f^n(c) < \alpha_n$.*

Proof:
Suppose that $f^n(c) \geq \alpha_n$. If $f^n(c) = \alpha_n$ then c is periodic by f^{2n} in contradiction to the hypothesis. Otherwise, f^{2n} maps c and α_n inside (α_n, c). Recall here that f^n is orientation-reversing on (α_n, c). Taking the limit as m goes to infinity of $f^{2nm}(c)$ we get a fixed point of f^{2n} which cannot be repelling.

$\square$

Fixed points of f^n. Let us suppose that n is a cutting time. Between α_n and c, f^n is orientation-reversing. So, there is a unique fixed point which will be called π_n. Then, we have a lemma:

Lemma 3.2.7 *If $n > 1$ is a proper cutting time, then $f^n(\alpha_{n'}) < \alpha_{n'}$.*

Proof:
Observe that $f^n(\alpha_{n'}) = f^{n-n'}(c)$. By Lemmas 3.2.4 and 3.2.5, $f^{n-n'}(c) < \alpha_{n-n'}$. Since n' is the cutting time directly preceding n, and $n - n'$ is also a cutting time by Lemma 3.2.4, $n - n' \leq n'$, hence by Lemma 3.2.2, $\alpha_{n'} \geq \alpha_{n-n'}$ and the claim of the present Lemma follows.

$\square$

It follows from Lemma 3.2.3 that f^n is monotone and increasing on $(\alpha_{n'}, \alpha_n)$. Hence, provided that n is proper, it has a unique fixed point χ_n on this interval. Points π_n and χ_n are the two periodic points of f^n closest to c. In the situation of Theorem 3.1, they are images by f of two fixed points of f^n in the closure of the restrictive interval.

Next, we consider the ordering of the sequence of points χ_n, α_n and π_n. For the same n, it is clearly increasing. We will show that the triples $\{\chi_n, \alpha_n, \pi_n\}$ are ordered monotonically.

Lemma 3.2.8 *Suppose that $n, \bar{n}$ are proper cutting times and $n > \bar{n}$. Then $\pi_{\bar{n}} < \alpha_n$.*

Proof:
Let us first observe that α_n and $\pi_{\bar{n}}$ are not equal, since one contains c in its forward orbit while the other does not. To get contradiction, assume $\alpha_n < \pi_{\bar{n}}$. Then $f^{\bar{n}}$ maps α_n into $(\pi_{\bar{n}}, c)$, hence inside (α_n, c). But then f^{n-1} is monotone in a neighborhood of $f^{\bar{n}}(\alpha_n)$. Hence $f^{\bar{n}+n-1}$ is monotone in a neighborhood of α_n which is a contradiction to the fact that α_n is a turning point of f^n.

$\square$

Lemma 3.2.9 *If $n, \bar{n}$ are proper cutting times and $\bar{n} < n$, then $\pi_{\bar{n}} \leq \chi_n$.*

Proof:
Suppose that $\pi_{\bar{n}} > \chi_n$ and see which way $f^{n+\bar{n}}$ will move the point $\pi_{\bar{n}}$. By Lemma 3.2.8, $\pi_{\bar{n}}$ is between $\alpha_{\bar{n}}$ and α_n, in the region where f^n is increasing. And so $\pi_{\bar{n}}$ will move right by f^n. Since $f^{\bar{n}}$ is decreasing on the right side of $\pi_{\bar{n}}$, f^n followed by $f^{\bar{n}}$ moves $\pi_{\bar{n}}$ to the left. But $f^{\bar{n}}$ followed by f^n moves it to the right, contradiction.

$\square$

Look on the other side of c. If n is a cutting time, let us look at the nearest critical point of f^n on the right side of c. It may not exist, but if it does we will call it ω_n. Let n^ω be the number of iterations it takes to send ω_n to c for the first time.

Lemma 3.2.10 *If n is a cutting time and ω_n is well-defined, then $n - n^\omega$ is a cutting time and $f^{n^\omega}(\alpha_n) = \alpha_{n-n^\omega}$.*

Proof:
Since f^n is monotone on (α_n, ω_n), f^{n-n^ω} is monotone on $(f^{n^\omega}(\alpha_n), c)$. By definition, the left endpoint of this interval goes to c, hence the condition of Definition 3.2.1 is fulfilled.

$\square$

Point $\hat{\chi}$. Suppose that n is a proper cutting time. Since $f^n(\omega_n)$ is the same as $f^{n-n^\omega}(c)$, Lemma 3.2.5 implies immediately that $f^n(\omega_n) < \alpha_{n-n^\omega} \leq \alpha_{n'}$. The last estimate uses Lemma 3.2.2. It means that somewhere on (α_n, ω_n) there is a unique point $\hat{\chi}_n$ which satisfies $f^n(\hat{\chi}_n) = \chi_n$. Note that $\hat{\chi}_n$ is well-defined by this condition even if ω_n is not. In this case, the monotone branch of f^n can be continued to the right until 1 is reached, and that means that the range of f^n extends to 0.

The renormalizable case. If n is the return time of a locally maximal restrictive interval I, then we already know by Lemma 3.2.1 that n is a cutting time. If the assumptions of Theorem 3.1 holds, n is also proper. The interval $(\chi_n, \hat{\chi}_n)$ is mapped by f^{n-1} monotonically onto I. Also, $\chi_n \leq f^n(c)$.

3.2.1 Reduction to a real estimate

Let us state the main real estimate. Recall the definitions from the paragraph on "fixed points of f^n" in the previous section, as they will be used throughout this section, too.

Proposition 3 *Let f be a unimodal quadratic polynomial. If n is a proper cutting time, and all orbits of f with period $2n$ and $4n$ are repelling, then*

$$\frac{c - f^n(c)}{c - \pi_n} \geq \rho$$

where ρ is a constant greater than $(1.3)^2$.

The estimate of Proposition 3 first appeared in a recent work of [33], with a slightly smaller ρ. A stronger version, with ρ as large as 1.8 was proved in [36]. A closely related estimate can be found in [26].

Our ρ is somewhere in between, since we were looking for a simple proof. Also, the papers quoted here get similar estimate where f is not necessarily a polynomial, or has a degenerate critical point. We will postpone the proof of Proposition 3 and show how it implies Theorem 3.1.

Bounded distortion.

Lemma 3.2.11 *Suppose f, n and I satisfy the hypotheses of Theorem 3.1. Let $I = (-a, a)$. There is a positive constant κ so that a neighborhood of $(\chi_n, \hat{\chi}_n)$ is mapped by f^{n-1} monotonically onto $(-(1+\kappa)a, (1+\kappa)a)$.*

Proof:
Let us first remark that this is a well-known result, see [30] for a survey of previous works and a proof. We give an independent proof based only on Proposition 3. This proof is different from all earlier ones in one technical aspect. Namely, it does not use the "shortest interval trick".

We know that f^n maps $\alpha_{n'}$ to $f^{n-n'}(c)$ and ω_n to $f^{n-n^\omega}(c)$, see Lemmas 3.2.4 and 3.2.10. Also, f^{n-1} is monotone on $(\alpha_{n'}, \omega_n)$ by the definition of cutting times. From Lemma 3.2.9, either $\pi_N \le \chi_n$ for $N = n'$ or $N = n - n^\omega$. For the same two values of N, Proposition 3 gives

$$\frac{c - f^N(c)}{c - \chi_n} \ge \rho \,.$$

Hence, we can pick $\kappa := \sqrt{\rho} - 1$.

$\square$

The controlling ratio. Here is the fundamental inequality about the distortion of the cross-ratio **Poin** by diffeomorphisms with negative Schwarzian derivative.

If $a < b < c < d$, then define their *cross-ratio* **Poin** by

$$\mathbf{Poin}(a, b, c, d) := \frac{|d - a||b - c|}{|b - a||d - c|} \,.$$

and *cross-ratio* **Cr** by

$$\mathbf{Cr}(a, b, c, d) := \frac{|b - a||d - c|}{|c - a||d - b|} \,.$$

Fact 3.2.1 *Diffeomorphisms which have negative Schwarzian derivative expand the cross-ratio* **Poin**

$$\mathbf{Poin}(a, b, c, d) < \mathbf{Poin}(f(a), f(b), f(c), f(d))$$

and contract the cross-ratio **Cr**

$$\mathbf{Cr}(a, b, c, d) > \mathbf{Cr}(f(a), f(b), f(c), f(d))$$

Fix n as in the statement of Theorem 3.1. Let $\gamma_1 = \frac{|f^n(c) - c|}{|\chi_n - c|}$ and $\gamma_2 = \frac{|f^{2n}(c) - c|}{|\chi_n - c|}$.

Lemma 3.2.12 *If n and f satisfy the hypotheses of Theorem 3.1, then $\sqrt{\gamma_1} > \frac{1}{3}$.*

Proof:
Without loss of generality, assume that $\chi_n < f^n(c)$. Set $x^2 = \frac{|\chi_n - c|}{|f^{n-n'}(c) - c|}$. Lemma 3.2.7 asserts that $|c - f^n(\alpha_{n'})| > |c - \alpha_{n'}| > |c - \chi_n|$. Hence, $x < 1$ and $\mathbf{Cr}(f^n(\alpha_{n'}), \chi_n, f^n(c), c)$ is larger than $\mathbf{Cr}(\alpha_{n'}, \chi_n, f^n(c), c)$ which in turn, by contracting property of **Cr**, is larger than

$$\mathbf{Cr}(f^{n-1}(\alpha_{n'}), f^{n-1}(\chi_n), f^{2n-1}(c), f^{n-1}(c)). \qquad (3.1)$$

We recall that $f^n(\alpha_{n'}) = f^{n-n'}(c)$ and $n - n'$ is a cutting time. Also $\alpha_{n'} < \chi_n < f^n(c)$. Therefore,

$$\frac{|f^{n-n'}(c) - f^n(c)|}{|f^{n-n'}(c) - \chi_n|} = \frac{|f^{n-n'}(c) - c| - |f^n(c) - c|}{|f^{n-n'}(c) - c| - |\chi_n - c|} = \frac{1 - \gamma_1 x^2}{1 - x^2}$$

Since χ_n and c are on the opposite sides of α_n, points $f^{n-1}(c)$ and $f^{n-1}(\chi_n)$ are separated by 0. Also, since $f^n(c) < \alpha_n$ by Lemma 3.2.6, points $f^{n-1}(c) = f^n(0)$ and $f^{2n-1}(c) = f^{2n}(0)$ must lie on the opposite sides of 0.

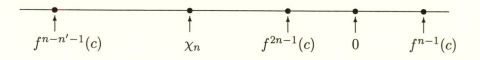

Moving $f^{2n-1}(c)$ to 0, we decrease the cross-ratio (3.1). Therefore,

$$\mathbf{Cr}(f^n(\alpha_{n'}), \chi_n, f^n(c), c) > \mathbf{Cr}(f^{n-1}(\alpha_{n'}), f^{n-1}(\chi_n), 0, f^{n-1}(c))$$

and by simple algebra, we get that γ_1 is larger than

$$\frac{1-\gamma_1 x^2}{1-x^2}\frac{\sqrt{|f^n(c)-c|}}{\sqrt{|\chi_n-c|}+\sqrt{|f^n(c)-c|}}\frac{\sqrt{|f^{n-n'}(c)-c|}-\sqrt{|\chi_n-c|}}{|\sqrt{|f^{n-n'}(c)-c|}}$$

$$=\frac{1-\gamma_1 x^2}{1-x^2}\frac{\sqrt{\gamma_1}}{1+\sqrt{\gamma_1}}(1-x)=\frac{\sqrt{\gamma_1}}{1+\sqrt{\gamma_1}}\frac{1-\gamma_1 x^2}{1+x}.$$

and

$$\sqrt{\gamma_1}(1+\sqrt{\gamma_1})\geq\frac{1-\gamma_1 x^2}{1+x}\geq\frac{1-\gamma_1}{2}. \qquad (3.2)$$

The last inequality holds since $x<1$.

We solve the quadratic inequality (3.2) with respect to $\sqrt{\gamma_1}$. That gives $\sqrt{\gamma_1}\geq\frac{1}{3}$.

$\square$

Lemma 3.2.13 *Assume the hypotheses of Theorem 3.1 and choose n and f. In addition, let*

$$\rho_n:=\frac{c-f^n(c)}{c-\pi_n}<2.$$

Then $\sqrt{\frac{\gamma_2}{\gamma_1}}\geq\frac{1}{3\sqrt{2}}$.

Proof:
We will consider two cases. If $f^{2n}(c)$ does not belong to the interval (π_n,c) then $|f^{2n}-c|\geq|\pi_n-c|$ and $\gamma_2/\gamma_1\geq\rho_n^{-1}\geq 1/2$, by the hypotheses of Lemma 3.2.13. If $f^{2n}(c)$ belongs to (π_n,c) then $f^{-1}(\pi_n,c]$ is a restrictive interval with return time $2n$. By the hypothesis of Theorem 3.1 all periodic points of periods $2n$ and $4n$ are repelling and we can apply Lemma 3.2.12 with $f^n(c)$ and χ_n replaced by $f^{2n}(c)$ and π_n respectively. We conclude that

$$\sqrt{\frac{|f^{2n}(c)-c|}{|\pi_n-c|}}\geq\frac{1}{3}$$

and that

$$\sqrt{\frac{\gamma_2}{\gamma_1}}\geq\sqrt{\frac{|c-\pi_n|}{c-f^n(c)}}\sqrt{\frac{|f^{2n}(c)-c|}{|\pi_n-c|}}\geq\frac{1}{3\sqrt{2}}.$$

$\square$

For the n from Theorem 3.1, we introduce a parameter $\theta :=$ $\frac{|c-\alpha_n|}{|c-f^n(c)|}$.

Lemma 3.2.14 *If f and n satisfy the hypotheses of Theorem 3.1, then*

$$\frac{\hat{\chi}_n - c}{c - \chi_n} \le \theta \cdot \frac{\sqrt{\gamma_1} - \gamma_1}{(1 - \rho^{-1/2})};.$$

The constant ρ comes from Proposition 3.

Proof:

If ω_n is well-defined, recall how ω_n is mapped to c by f^{n^ω} and that by Lemma 3.2.10 $\bar{n} := n - n^\omega$ is a cutting time. Applying f^{n-1} to points $\alpha_n, c, \hat{\chi}_n, \omega_n$ and using expansion property of **Poin**, we get

$$\frac{|f^{n-1}(\hat{\chi}_n) - f^{n-1}(c)|}{|\hat{\chi}_n - c|} \frac{|f^{\bar{n}-1}(c) - 0|}{|\alpha_n - \omega_n|} \ge$$
$$\frac{|f^{n-1}(c) - 0|}{|\alpha_n - c|} \frac{|f^{\bar{n}-1}(c) - f^{n-1}(\hat{\chi}_n)|}{|\hat{\chi}_n - \omega_n|}. \tag{3.3}$$

If ω_n is not well-defined, it can be moved to infinity, and the second factor on each side of the inequality drops out. Since $|\alpha_n - \omega_n| \ge |\hat{\chi}_n - \omega_n|$, we get

$$|\hat{\chi}_n - c| \le |\alpha_n - c| \frac{|f^{n-1}(\hat{\chi}_n) - f^{n-1}(c)|}{|f^{n-1}(c) - 0|} \frac{|f^{\bar{n}-1}(c) - 0|}{|f^{\bar{n}-1}(c) - f^{n-1}(\hat{\chi}_n)|} \tag{3.4}$$

where the last factor on the right side becomes 1 when ω_n is not well-defined.

Since the range of (α_n, c) covers 0 by virtue of n being a cutting time, points $f^{n-1}(c)$ and $f^{n-1}(\hat{\chi})$ are on the same side of 0. Hence,

$$\frac{|f^{n-1}(\hat{\chi}_n) - f^{n-1}(c)|}{|f^{n-1}(c) - 0|} = \sqrt{\frac{|\chi_n - c|}{|c - f^n(c)|}} - 1. \tag{3.5}$$

If ω_n is well-defined, point $f^{\bar{n}-1}(c) = f^{n-1}(\omega_n)$ is also on the same side and further away than $f^{n-1}(\hat{\chi}_n)$, hence

$$\frac{|f^{\bar{n}-1}(c) - f^{n-1}(\hat{\chi}_n)|}{|f^{\bar{n}-1}(c) - 0|} = 1 - \sqrt{\frac{|\chi_n - c|}{|f^{\bar{n}}(c) - c|}}.$$

By Lemma 3.2.9 and Proposition 3

$$\frac{|\chi_n - c|}{|f^{\bar{n}}(c) - c|} \le \frac{|\pi_{\bar{n}} - c|}{|f^{\bar{n}}(c) - c|} \le \rho^{-1} .$$

Hence

$$\frac{|f^{\bar{n}-1}(c) - f^{n-1}(\hat{\chi}_n)|}{|f^{\bar{n}-1}(c) - 0|} \ge 1 - \rho^{-1/2} .$$

Putting together this and equality (3.5) and substituting into estimate (3.4), we get

$$|\hat{\chi}_n - c| \le |\alpha_n - c| \frac{\sqrt{\frac{|\chi_n - c|}{|c - f^n(c)|}} - 1}{1 - \rho^{-1/2}} .$$

Note that this remains true even if ω_n is not well-defined. In that case we should have replaced the last factor in estimate (3.4) with 1, and we ended up replacing it with $\frac{1}{1 - \rho^{-1/2}}$.

Multiplying and dividing by the same factors, we arrive at

$$\frac{|\hat{\chi}_n - c|}{|\chi_n - c|} \le \frac{|f^n(c) - c|}{|\chi_n - c|} \frac{|\alpha_n - c|}{|f^n(c) - c|} \frac{\sqrt{\frac{|\chi_n - c|}{|c - f^n(c)|}} - 1}{1 - \rho^{-1/2}} . \qquad (3.6)$$

Note that the second factor on the right side of estimate (3.6) is no greater than 1, by Lemma 3.2.6 and the hypothesis of Theorem 3.1 about fixed points of f^{2n}.

After substituting γ_1, estimate (3.6) gives the claim of the lemma.

$\square$

By Lemma 3.2.6 and the hypothesis of Theorem 3.1 about fixed points of f^{2n} we have a trivial estimate that $\theta < 1$. Here is a better estimate.

Lemma 3.2.15 *Suppose that n and f satisfy the hypothesis of Theorem 3.1. Then,*

$$\theta \le \frac{1 + \sqrt{\rho^{-1}}\sqrt{\gamma_2}}{1 + \sqrt{\frac{\gamma_2}{\gamma_1}}} .$$

Proof:
Assume first that ω_n is well defined. We check directly that

$$\theta \leq \frac{|\alpha_n - c|}{|f^n(c) - c|} \frac{|f^n(c) - \omega_n|}{|\alpha_n - \omega_n|} = \frac{\textbf{Poin}(f^n(c), \alpha_n, c, \omega_n)}{\textbf{Cr}(f^n(c), \alpha_n, c, \omega_n)} .$$

We want to apply f^{n-1} to the quadruple $\{f^n(c), \alpha_n, c, \omega_n\}$. By Fact 3.2.1, cross-ratio $\textbf{Poin}/\textbf{Cr}$ is expanded by iterates of f. Therefore,

$$\theta \leq \frac{|f^{n-1}(c) - 0|}{|f^{2n-1}(c) - f^{n-1}(c)|} \frac{|f^{2n-1}(c) - f^{n-1}(\omega_n)|}{|0 - f^{n-1}(\omega_n)|}$$

By Lemma 3.2.10, $\bar{n} := n - n^\omega$ is a cutting time. Since $f^{n-1}(\omega_n)$ and $f^{n-1}(c)$ lie on the other side of 0 than $f^{2n-1}(c)$, we get that

$$\begin{aligned}
\theta \;&\leq\; \frac{\sqrt{\gamma_1}}{\sqrt{\gamma_1} + \sqrt{\gamma_2}} \frac{\sqrt{|f^{2n}(c) - c|} + \sqrt{|c - f^{\bar{n}}(c)|}}{\sqrt{|f^{\bar{n}}(c) - c|}} \\
&<\; \frac{1}{1 + \sqrt{\frac{\gamma_2}{\gamma_1}}} \left(1 + \sqrt{\gamma_1}\sqrt{\frac{\gamma_2}{\gamma_1}} \frac{\sqrt{|c - \chi_n|}}{\sqrt{|c - f^{\bar{n}}(c)|}}\right) \\
&\leq\; \frac{1 + \sqrt{\rho^{-1}}\sqrt{\gamma_1}\sqrt{\frac{\gamma_2}{\gamma_1}}}{1 + \sqrt{\frac{\gamma_2}{\gamma_1}}}
\end{aligned}$$

where the last inequality uses the estimate

$$\frac{c - \chi_n}{c - f^{\bar{n}}(c)} \leq \frac{c - \pi_{\bar{n}}}{c - f^{\bar{n}}(c)} \leq \rho^{-1} .$$

Here, the first step is based on Lemma 3.2.9, namely that $\pi_{\bar{n}} < \chi_n < f^n(c)$ and the second follows from Proposition 3.

If ω_n is not well defined then the second factor on the right-hand side of (3.7) drops out and we obtain an even stronger estimate.

$\square$

Lemma 3.2.16 *Assume Proposition 3 with $\sqrt{\rho} = 1.3$. Under the hypotheses of Theorem 3.1, there exists a positive constant $\Omega < 1$ so that*

$$\theta \cdot \frac{\sqrt{\gamma_1} - \gamma_1}{(1 - \rho_n^{-1/2})} < \Omega .$$

Here,

$$\rho_n := \frac{c - f^n(c)}{c - \pi_n} \; .$$

Proof:
First consider the case when $\rho_n \geq 2$. The maximum of $\sqrt{\gamma_1} - \gamma_1$ occurs when $\gamma = \frac{1}{4}$ and equals $\frac{1}{4}$. The parameter θ is less than 1. If $\rho_n \geq 2$ then $\frac{1}{4(1-\rho_n^{-1/2})} = \frac{2+\sqrt{2}}{4} < 1$ which proves Lemma 3.2.16. Suppose that $\rho_n < 2$. Set $\eta := \sqrt{\frac{\gamma_2}{\gamma_1}}\rho_n^{-1/2}$ and $s := \sqrt{\gamma_1}$ and substitute them to the estimate on θ given by the claim of Lemma 3.2.15

$$\theta \leq \frac{1 + \eta s}{1 + \eta\sqrt{\rho_n}}$$

and the claim of Lemma 3.2.16 becomes

$$\frac{s(1-s)}{(1 - \rho_n^{-1/2})} \frac{1 + \eta s}{1 + \eta\sqrt{\rho_n}} < \Omega < 1. \tag{3.7}$$

First estimate the maximum of $g_\eta(s) := s(1-s)(1+\eta s)$ with respect to s when η is fixed. The derivative is

$$\frac{dg_\eta}{ds} = \eta s(2 - 3s) + 1 - 2s \; .$$

This is negative if $s > 2/3$, hence the maximum occurs for $s_\eta \leq 2/3$. Since $s(1 - s) \leq \frac{1}{4}$, we obtain

$$g_\eta(s) \leq \frac{1}{4}(1 + \frac{2}{3}\eta) \; .$$

Consequently,

$$\frac{s(1-s)}{(1 - \rho_n^{-1/2})} \frac{1 + \eta s}{1 + \eta\sqrt{\rho_n}} \leq \frac{1 + \frac{2}{3}\eta}{4(1 - \rho_n^{-1/2})(1 + \eta\sqrt{\rho_n})} \; .$$

The right side of this inequality is an increasing function of η since $\rho_n > 1$. From Lemma 3.2.13,

$$\eta = \sqrt{\gamma_2}\gamma_1\rho_n^{-1/2} \geq \frac{3}{\sqrt{2\rho_n}} \geq \frac{1}{6}$$

because of our standing assumption that $\rho_n < 2$. Hence, we can substitute $\eta = \frac{1}{6}$ which leads to

$$\frac{s(1-s)}{(1-\rho_n^{-1/2})}\frac{1+\eta s}{1+\eta\sqrt{\rho_n}} \leq \frac{6+\frac{2}{3}}{4(5+\sqrt{\rho_n}-\frac{6}{\sqrt{\rho_n}})}$$

which is decreasing with ρ_n. The value of the right side for $\sqrt{\rho_n} :=$ 1.3 is less than 1 and so can serve as the Ω.

$\square$

Proof of Theorem 3.1. Let I, β, f and n be as in the hypotheses of Theorem 3.1. For $\epsilon \geq 0$, let $\chi(\epsilon)$ be defined as the largest number which satisfies the conditions that $|f^{n-1}(\chi(\epsilon))| = (1+\epsilon)\frac{|I|}{2}$ and $\chi(\epsilon) \leq \chi_n$. Then $\hat{\chi}(\epsilon)$ is the point on the other side of c so that $|f^{n-1}(\hat{\chi}(\epsilon))| = |f^{n-1}(\chi(\epsilon))|$ and f^{n-1} is monotone on $(\chi(\epsilon), \hat{\chi}(\epsilon))$. By Lemma 3.2.11, $\hat{\chi}(\epsilon)$ is well-defined for $\epsilon \leq \kappa$. Also, $\chi(0) = \chi_n$ and $\hat{\chi}(0) = \hat{\chi}_n$.

Using the real Köbe lemma (Fact 2.1.1) and Lemma 3.2.11, find $0 < \epsilon_1 < \kappa$ so that $|(f^{n-1})'(x)| \geq \frac{|(f^{n-1})'(\chi_n)|}{\sqrt{\beta}}$ for $x \in (\chi(\epsilon_1), \chi_n]$. Indeed, if ϵ_1 is small compared with the κ from Lemma 3.2.11 and 1, the extendibility of f^{n-1} from $(\chi(\epsilon_1), \chi_n)$ goes to 1. Also by Fact 2.1.1 applied to f^{n-1} on $(\chi(\epsilon_1), \hat{\chi}(\epsilon_1))$,

$$|(f^{n-1})'(x)| \geq K_1\frac{|I|}{|\hat{\chi}_n - \chi_n|}$$

for all $x \in (\chi(\epsilon_1), \hat{\chi}(\epsilon_1))$. Numbers ϵ_1 and K_1 depend only on κ and β.

Since $|(f^n)'| \geq \beta$ at the endpoints of I, and the derivative of the quadratic function increases away from 0, we have that

$$|(f^n)'(x)| \geq \sqrt{\beta}$$

for $x \in f^{-1}((\chi(\epsilon_1), \chi_n])$. Thus, for any $0 < \epsilon \leq \epsilon_1$,

$$|f^n(y)| = (1+\epsilon)\frac{|I|}{2}$$

for some $|y| \leq (1+\epsilon/\sqrt{\beta})\frac{|I|}{2}$. Also, for the same range of ϵ, the Hausdorff distance $(\chi(\epsilon), \hat{\chi}(\epsilon))$ and $(\chi_n, \hat{\chi}_n)$ is bounded by $K_1\epsilon(\hat{\chi}_n -$

χ_n). By Lemmas 3.2.14 and 3.2.16, we can choose the $0 < \epsilon_2 \leq \epsilon_1$ which depends only on Ω, K_1 and ϵ so that

$$\frac{\hat{\chi}(\epsilon_2) - c}{c - \chi(\epsilon_2)} < 1 . \tag{3.8}$$

Now consider the geodesic neighborhood (disk)

$$V := D\left(\frac{\pi}{2}, f^{n-1}(\chi(\epsilon_2), \hat{\chi}(\epsilon_2))\right) .$$

By Fact 2.1.2 there is a

$$V' \subset D\left(\frac{\pi}{2}, \chi(\epsilon_2), \hat{\chi}(\epsilon_2)\right)$$

so that f^{n-1} maps V' onto V univalently. Finally, by estimate (3.8), $U := f^{-1}(V')$ is contained in the disk with $f^{-1}(\chi(\epsilon_2), c]$ as the diameter. It is clear from the construction that f^n maps U onto V in a proper fashion with degree 2. Since the diameter of V is $(1 + \epsilon_2)|I|$ and the diameter of U is no more than $(1 + \frac{\epsilon_2}{\sqrt{\beta}})|I|$,

$$\mathrm{mod}\,(V \setminus \overline{U}) \geq \log \frac{1 + \epsilon_2}{1 + \epsilon_2/\sqrt{\beta}} .$$

This proves Theorem 3.1.

3.2.2   Proof of the real estimate

We will adopt a notation

$$\rho_n := \frac{c - f^n(c)}{c - \pi_n}$$

where it is understood that n is a cutting time so that π_n is well-defined.

Lemma 3.2.17 *Suppose that n is a proper cutting time, ω_n is well-defined, and all orbits of period $2n$ are repelling.*

Let $0 < \overline{n} < n$ be defined by $f^{n-\overline{n}}(\omega_n) = c$. Notice that $\overline{n}$ is a cutting time by Lemma 3.2.10. Suppose that $\rho_n < \rho_{\overline{n}}$ and $\rho_n \leq 2$. Then

$$(\rho_n - 1)(\sqrt{\rho_n} - 1) \geq (1 - \frac{1}{\sqrt{\rho_{\overline{n}}}}) .$$

Proof:

We apply f^{n-1} to points $\alpha_n, \pi_n, c, \omega_n$ and use the expansion property of **Poin**. The result is

$$\frac{|f^{n-1}(c) - f^{n-1}(\pi_n)|}{|c - \pi_n|} \frac{|f^{\overline{n}-1}(c) - 0|}{|\omega_n - \alpha_n|} \geq \frac{|f^{\overline{n}-1}(c) - f^{n-1}(c)|}{|\omega_n - c|} \frac{|f^{n-1}(\pi_n) - 0|}{|\alpha_n - \pi_n|}.$$

Since $|\omega_n - \alpha_n| \geq |\omega_n - c|$, this leads to

$$\frac{|f^{n-1}(c) - f^{n-1}(\pi_n)|}{|f^{n-1}(\pi_n) - 0|} \geq \frac{|f^{\overline{n}-1}(c) - f^{n-1}(c)|}{|f^{\overline{n}-1}(c) - 0|} \frac{|f^{\overline{n}-1}(c) - f^{n-1}(c)|}{|\alpha_n - \pi_n|}.$$
$$(3.9)$$

As points $f^{n-1}(c)$ and $f^{n-1}(\pi_n)$ are on the same side of 0, we get

$$\frac{|f^{n-1}(c) - f^{n-1}(\pi_n)|}{|f^{n-1}(\pi_n) - 0|} = \sqrt{\frac{|f^n(c) - c|}{|\pi_n - c|}} - 1 = \sqrt{\rho_n} - 1.$$

Point $f^{n-1}(c)$ is between $f^{\overline{n}-1}(c) = f^{n-1}(\omega_n)$ and 0, hence the first fraction on the right side of (3.9) can be rewritten as

$$\frac{|f^{\overline{n}-1}(c) - f^{n-1}(c)|}{|f^{\overline{n}-1}(c) - 0|} = 1 - \sqrt{\frac{|f^n(c) - c|}{|f^{\overline{n}}(c) - c|}}.$$

Estimate (3.9) becomes

$$\sqrt{\rho_n} - 1 \geq (1 - \sqrt{\frac{|f^n(c) - c|}{|f^{\overline{n}}(c) - c|}} \frac{|\pi_n - c|}{|\alpha_n - \pi_n|}.$$
$$(3.10)$$

Finally, we use

$$\frac{|f^n(c) - c|}{|f^{\overline{n}}(c) - c|} = \frac{\rho_n}{\rho_{\overline{n}}} \frac{|\pi_n - c|}{|\pi_{\overline{n}} - c|}$$

to arrive at

$$\sqrt{\rho_n} - 1 \geq \left(1 - \sqrt{\frac{\rho_n}{\rho_{\overline{n}}}} \sqrt{\frac{|\pi_n - c|}{|\pi_{\overline{n}} - c|}}\right) \frac{|\pi_n - c|}{|\alpha_n - \pi_n|}.$$
$$(3.11)$$

Now the argument splits into two cases.

Case I. We assume that $|f^n(c) - c| \leq |\pi_{\overline{n}} - c|$.
By Lemma 3.2.17, $|f^n(c) - \pi_n| \geq |\alpha_n - \pi_n|$ and we can write

$$\sqrt{\rho_n} - 1 \geq \left(1 - \sqrt{\frac{\rho_n}{\rho_{\overline{n}}}}\sqrt{\frac{|\pi_n - c|}{|f^n(c) - c|}}\right)\frac{|\pi_n - c|}{|\alpha_n - \pi_n|}$$

instead of estimate (3.11), and that is equivalent to

$$\sqrt{\rho_n} - 1 \geq \left(1 - \sqrt{\frac{1}{\rho_{\overline{n}}}}\right)\frac{|\pi_n - c|}{|\alpha_n - \pi_n|}. \qquad (3.12)$$

So, the hypothesis of this case gives

$$\sqrt{\rho_n} - 1 \geq \left(1 - \sqrt{\frac{1}{\rho_{\overline{n}}}}\right)\frac{|\pi_n - c|}{|f^n(c) - \pi_n|}.$$

Since

$$\frac{|f^n(c) - \pi_n|}{|\pi_n - c|} = \rho_n - 1$$

the claim of Lemma 3.2.17 follows.

Case II. We assume that $|\pi_{\overline{n}} - c| < |f^n(c) - c|$.
By Lemma 3.2.9, $|\pi_{\overline{n}} - \pi_n| \geq |\chi_n - \alpha_n|$. Estimate (3.11) becomes

$$\sqrt{\rho_n} - 1 \geq \left(1 - \sqrt{\rho_n \rho_{\overline{n}}}\sqrt{\frac{|\pi_n - c|}{|\pi_{\overline{n}} - c|}}\right)\frac{|\pi_n - c|}{|\pi_{\overline{n}} - \pi_n|}.$$

Since

$$\frac{|\pi_n - c|}{|\pi_{\overline{n}} - \pi_n|} = \frac{|\overline{\pi}_n - c|}{|\pi_{\overline{n}} - \pi_n|} - 1$$

we get

$$\sqrt{\rho_n} \geq \left[-\sqrt{\frac{\rho_n}{\rho_{\overline{n}}}\frac{|\pi_n - c|}{|\pi_{\overline{n}} - c|}}\frac{|\pi_n - c|}{|\pi_{\overline{n}} - \pi_n|} + \frac{|\overline{\pi}_n - c|}{|\pi_{\overline{n}} - \pi_n|}\right].$$

Multiplying both sides by $\frac{|\pi_{\overline{n}} - \pi_n|}{|\pi_{\overline{n}} - c|}$ we arrive at

$$\sqrt{\rho_n}\frac{|\pi_{\overline{n}} - \pi_n|}{|\pi_{\overline{n}} - c|} \geq \left[-\sqrt{\frac{\rho_n}{\rho_{\overline{n}}}\frac{|\pi_n - c|}{|\pi_{\overline{n}} - c|}}\frac{|\pi_n - c|}{|\pi_{\overline{n}} - c|} + 1\right]. \qquad (3.13)$$

To facilitate the notation, let us abbreviate

$$\zeta := \frac{|\pi_n - c|}{|\pi_{\overline{n}} - c|}$$

and estimate (3.13) can be rewritten as

$$\sqrt{\rho_n}(1 - \zeta) - \left[1 - \sqrt{\frac{\rho_n}{\rho_{\overline{n}}}} \zeta^{3/2}\right] \geq 0 \, .$$

Solving with respect to ρ_n we arrive at

$$\sqrt{\rho_n} \geq \frac{\sqrt{\rho_{\overline{n}}}}{(1 - \zeta)\rho_{\overline{n}} + \zeta^{3/2}} \, . \tag{3.14}$$

It is time to use the hypothesis of this case, i.e. $|\pi_{\overline{n}} - c| \leq |f^n(c) - c|$. It means $\zeta \geq \rho_n^{-1}$. On the other hand, clearly $\zeta \leq 1$. We will look for the minimum of the left-hand side of (3.14) with respect to ζ subject to those conditions. Let us examine the sign of the derivative of the right-hand side. It is the same as the sign of

$$\rho_{\overline{n}} - \frac{3}{2}\sqrt{\zeta}\sqrt{\rho_n}.$$

This expression is decreasing with ζ, so that even if the derivative vanishes, this point must be a local maximum. It develops that the minimum is taken either at $\zeta = 1$ or $\zeta = \rho_n^{-1}$.

When the first value is substituted into estimate (3.14), it yields $\rho_n \geq \rho_{\overline{n}}$. So this is ruled out by the hypothesis of Lemma 3.2.17. The second possibility leads to

$$\sqrt{\rho_n} \geq \frac{\rho_n \sqrt{\rho_{\overline{n}}}}{\sqrt{\rho_{\overline{n}}}(\rho_n - 1) + (\rho_n)^{-1/2}} \, .$$

Getting rid of the denominator, one gets

$$\sqrt{\rho_n}\sqrt{\rho_{\overline{n}}}(\rho_n - 1) + 1 \geq \rho_n \sqrt{\rho_{\overline{n}}} \, .$$

Next, we divide both sides by $\sqrt{\rho_{\overline{n}}}$ and group the terms which include ρ_n on the left side:

$$\rho_n^{3/2} - \sqrt{\rho_n} - \sqrt{\rho_n} \geq -\rho_{\overline{n}}^{-1/2} \, .$$

Adding 1 to both sides, we end up with

$$(\rho_n - 1)(\sqrt{\rho_n} - 1) \geq (1 - \rho_{\overline{n}}^{-1/2})$$

which gives the claim of Lemma 3.2.17.

$$\square$$

An inductive estimate for ρ_n**.** The function $\sqrt{x}(x-1)$ is increasing for $x > 0$. Hence, the equation

$$\sqrt{x}(x-1) = 1$$

has precisely one solution. Denote this solution by ρ_0.

Lemma 3.2.18 *Suppose that n is a proper cutting time and all orbits of period $2n$ are repelling. Then ρ_n is at least equal to the minimum of 2 and ρ_0.*

Proof:
We will proceed by contradiction. For a given f, let n be the smallest cutting time for which the claim of the Lemma fails. We will show that the hypotheses of Lemma 3.2.17 are satisfied.

First, we have to check that ω_n is well-defined. Otherwise, the cross-ratio property applied to $\alpha_n, \pi_n, c, \infty$ gives

$$\frac{\pi_n - f^n(c)}{c - \pi_n} \geq \frac{\pi_n - \alpha_n}{c - \pi_n} \geq \frac{c - \pi_n}{\pi_n - f^n(c)}.$$

The first inequality follows from Lemma 3.2.6. Hence,

$$\pi_n - f^n(c) \geq c - \pi_n$$

which means that $\rho_n \geq 2$, contrary to our assumption that the claim of Lemma 3.2.18 has failed.

So, the cutting time $\bar{n}$ is well defined (see the statement of Lemma 3.2.17). Next, $\rho_n < \rho_{\bar{n}}$ since n was the first cutting time for which $\rho_n < \min(2, \rho_0)$.

Hence, we have

$$(\rho_n - 1)(\sqrt{\rho_n} - 1) \geq (1 - \frac{1}{\sqrt{\rho_{\bar{n}}}}) \geq (1 - \frac{1}{\sqrt{\rho_n}})$$

and $\sqrt{\rho_n}(\rho_n - 1) \geq 1$. Thus, $\rho_n \geq \rho_0$ contrary to our assumption.

$\square$

Proof of the Proposition. We have that $\rho_n \geq \min(\rho_0, 2)$. By a direct check $\sqrt{(1.3)^2}(1-(1.3)^2) = 1.3 \cdot 0.3 \cdot 2.3 < 1$, hence $\rho_0 > (1.3)^2$ and Proposition 3 has been proved.

Chapter 4

Linear Growth of Moduli

4.1   Box Maps and Separation Symbols

4.1.1   A general outline

Box mappings were introduced in [13] as a tool for studying the dynamics of real unimodal polynomials. In the same paper, the main property of growing moduli was proved. This generalized earlier results obtained for certain ratios on the real line. In [14], a more general result was presented with a slightly different proof, not more complicated that the original proof of a weaker result in [13]. We state the main theorem of [14] as Theorem 1.2. The generalization consists in allowing a large class of holomorphic box mappings without any connection with real dynamics. Theorem 1.2 found already applications in the recent work of Przytycki, [35], about the equality between the Hausdorff dimension of Julia sets of complex quadratic polynomials and their hyperbolic Hausdorff dimension, and Graczyk and Smirnov, [12], about the Hausdorff dimension of the Julia set for the quadratic Fibonacci map. Also, a recent work of Heckman concerning bimodal polynomials relies on our Theorem 1.2, see [19].

The proof closely follows that of [14] and relies consistently on "conformal roughness" of curves separating certain dynamically defined annuli. This gives rise to the linear growth of moduli claimed in Theorem 1.2.

One cannot help feeling certain dissatisfaction at not being able to state this important property of quadratic dynamics in simple

terms. The statement has to involve the inducing procedure as the
annuli we are discussing are only constructed inductively in this process.

Organization of the chapter. The chapter has three sections.
In the first, the basic concepts of geometry of box mappings are presented including the separation index. The separation index and the
inducing process are the same as in [13]. The weak monotonicity of
the separation index with respect to the inducing process is shown
together with some cases of actual growth. Our reasoning apart from
a few applications of Lemma 4.1.4 is almost purely combinatorial.
The proof of the analytical Lemma 4.1.4 is postponed till the end
of the second section. Section 4.2 has little to do with dynamical
systems and can be read independently. It deals with some conse-
quences of the classical theory of modulus due to Teichmüller. We
work with the notion of *conformal roughness* of Jordan curves, already introduced in [13]. The last section of the chapter is devoted
to deriving the Main Theorem about the linear growth of moduli.
The conformal roughness of boundaries of boxes plays an important
role here.

Let us now recall the box inducing process from Section 1.3.1.

Filling-in. Suppose that φ is a type II box mapping which yields a
type I box mapping ϕ in the process of filling-in. Recall that filling-in
means replacing the φ on the union of its univalent domains by the
first entry map into the central domain B. That is, $\phi(z) := \varphi^{n(z)}(z)$
where

$$n(z) = \min\{n : \varphi^n(z) \in B\} .$$

If $n(z)$ is infinite, $\phi(z)$ is undefined.

This implies a certain combinatorial structure on the branches of
ϕ. Each branch ζ of ϕ is a restriction of

$$\bar{\zeta} = \zeta_n \circ \cdots \circ \zeta_1$$

where ζ_i are branches of φ. In that context, ζ_1 is called the *parent branch* of ζ, and the domain of ζ_1 is called the *parent domain*.
Certainly, the domain of ζ is compactly contained in its parent domain. Notice also that $\bar{\zeta}$ naturally maps onto B' even though ζ by

definition maps onto B. Hence, every univalent branch of a type I holomorphic box mapping arising from a type II holomorphic box mapping has a univalent *dynamical extension* onto B'. If we pick another branch of ϕ, say η, it may be that $\overline{\eta} = \zeta_{n+k} \circ \cdots \circ \zeta_1$ or that $\overline{\eta} = \zeta_{n-k} \circ \cdots \circ \zeta_1$. In the first case, we say that η is *subordinate* to ζ, in the second case ζ is *subordinate* to η, and in the remaining case we will say that they are *independent*.

Every parent domain contains exactly one branch of ϕ which is just the restriction of the parent branch. This branch will be called *maximal*. This is the same as maximality in the usual sense with respect to the partial ordering of subordination.

Critical filling. Now assume that a type I box mapping ϕ gives rise to a type II box mapping φ in a step of critical filling. Recall that this involves first constructing the map ϕ_0 which is the same as ϕ except on the central domain where the central branch ψ of ϕ was replaced by the identity map. Then $\varphi = \phi_0 \circ \phi$. The branches of φ are referred to as *parent branches*. This is consistent with the terminology of the preceding fragment, since these branches will be filled-in in the course of the next inducing step. If ϕ makes a non-close return, among the parent branches there are two special ones called *immediate*. Both are characterized as being restrictions of the central branch of ϕ to subdomains mapped on the central domain. Each of these immediate parent branches gives rise to single maximal branch of the type I mapping $\tilde{\phi}$ which is eventually derived from φ by filling-in. These are called *immediate branches* of $\tilde{\phi}$.

Symmetric box mappings. A holomorphic box mapping is *symmetric* if B is symmetric with respect to 0 and the central branch can be factored as $H(z^2)$ where H is univalent. Among symmetric box mappings, we consider *type I* and *type II* box mappings. We recall that type I box mapping is determined by the condition that all non-central branches have range B. Similarly, type II is characterized by the property that all non-central branches have range B'. In this section we deal exclusively with symmetric Jordan box mappings either of type I or type II.

4.1.2   The growth of moduli

We recall Theorem 1.2 which will be proved in this chapter.

Theorem 1.2 *Let ϕ be a symmetric Jordan type II box mapping, and let B and B' denote the domain and range of its central branch, respectively. Let ϕ_0 be the box mapping obtained from ϕ by filling-in and ϕ_i form a sequence, finite or not, of holomorphic box mappings set up so that ϕ_{i+1} is derived from ϕ_i by the type I inducing step for $i \geq 0$. Suppose that for ϕ, $\mathrm{mod}\,(B' \setminus B) \geq \alpha_0$. For every $\alpha_0 > 0$ there is a number $C > 0$ with the property that for every i*

$$\mathrm{mod}\,(B_i' \setminus B_i) \geq C \cdot i \,.$$

In particular, for every i we have that every connected component of the domain of ϕ_i contained in B_i' is separated from the complement of B_i' by an annulus with modulus at least $C \cdot i$.

Generalizations.   Sometimes one may want to have Theorem 1.2 without the restriction about the mapping being symmetric or Jordan. These versions can be derived without much difficulty from Theorem 1.2. For details, see [14].

Extendibility of branches.   As we have already observed a univalent branch of type I holomorphic box mapping has a canonical extension mapping over B'. The central branch ψ of a symmetric box mapping can be factored as $h(z^2)$ and h ranges over B'. Now, an analytic extension of ψ will be defined as composition of z^2 with a univalent continuation of h.

Fact 4.1.1 *Let ϕ be a type I holomorphic box mapping. Let ϕ_i be obtained from ϕ by a series of i simple inducing steps all showing close returns. Then every univalent branch of ϕ_i has a univalent extension up to the range B' and the domain of such an extension is contained in $B_i' \setminus B_i$.*

Proof:

We conduct the proof by induction with respect to i. For $i = 0$ the needed extension is just the canonical extension, so the claim is true. Then, notice that all univalent branches of ϕ_i are compositions of parent branches. Parent branches, in turn, are all in the form $p \circ \psi$ where ψ is the central branch of ϕ and p range over the set of univalent branches of some ϕ_{i-1}. Hence each parent branch has an extension with the range B' defined as ψ followed by the extension of p which exists by the hypothesis of induction. The induction step follows.

$\square$

Strategy of the proof. The problem with the growth of mod $(B_i' \setminus B_i)$ is that it is not monotonic. So the first thing we do is to introduce another quantity for type I box mappings, called separation index, which is monotonic with respect to a type I inducing step and bounds mod $(B_i' \setminus B_i)$ from below up to a multiplicative constant. For the separation index, it is easy to show that it is non-decreasing. The difficulty is in showing an actual increase after a bounded number of type I inducing steps. This is derived from the classical theory of modulus from the Thirties. Our main tool is the Modulsatz due to Teichmüller.

4.1.3   Separation symbols

Whenever we talk about separation symbols for a type I box mapping, we tacitly assume that it was derived by filling-in from a type II box mapping.

Conformal moduli. Some of the facts about conformal moduli used in our proofs are very simple indeed. By an annulus, synonymous with "ring domain" we mean any open region of the plane homeomorphic to the punctured plane. For simplicity of notation, we will often not distinguish between annuli delimited by Jordan curves and their closures. We adopt the following convention while calculating moduli. If B contains a Jordan annulus A and is contained in its closure $\overline{A}$ then we set $\operatorname{mod} B := \operatorname{mod} A$.

Canonical mapping. A classical theorem says that every annulus A can be mapped conformally onto the ring $\{z : 0 \le d < |z| <$

1} where d is unique and $-\log d$ is called the *modulus* of the annulus. We often invoke the inverse of this map and call it the *canonical mapping* of A. The modulus is a conformal invariant, so d depends only on A.

Modulus and holomorphic covers. Let A and A' be annuli and $f : A \to A'$ be a holomorphic cover of degree k. Then $\operatorname{mod} A' = k \cdot \operatorname{mod} A$. This is also a straightforward consequence of the existence of the canonical map. By the canonical maps, f can be lifted to a cover g of one round ring onto another. The covering map g can be continued to a conformal map of the punctured plane onto itself by a sequence of reflections, hence g is just $z \to e^{it} z^k$ and the claim follows.

Super-additivity. We will say that two annuli C_1 and C_2 are *nesting* provided that they are disjoint and one fits into the bounded connected component of the complement of the other one. Then we define a commutative operation $C_1 \oplus C_2$ on pairs of nesting annuli which results into the smallest annulus containing both C_1 and C_2.

The *super-additivity* of the modulus means that

$$\operatorname{mod}(C_1 \oplus C_2) \geq \operatorname{mod} C_1 + \operatorname{mod} C_2 .$$

This follows from Lemma 6.3 on page 35 in [25]. In the future, we will state Teichmüller's Modulsatz which gives stringent conditions under the equality can be attained or approximately attained.

Another estimate. The final estimate concerns the situation when a mapping from one annulus onto another is holomorphic and proper but not a cover (has critical points). Figure 4.1 illustrates the setting of Lemma 4.1.1.

Lemma 4.1.1 *Consider a topological disk D_1', an annulus $U_1 \subset D_1'$, and the topological disk D_1 determined as the union of U_1 with the bounded connected component of its complement. Denote $W_1 := D_1' \setminus \overline{D_1}$. Then look at a holomorphic mapping f in the form $h(z^2)$ with h univalent, f from D_1' onto a topological disk D_2', proper of degree 2, and assume that f is univalent in D_1. Define $D_2 := f(D_1)$,*

$W_2 := D'_2 \setminus \overline{D_2}$ and $U_2 := f(U_1)$. *Choose non-negative numbers* σ_1 *and* σ_2 *so that*

$$\sigma_2 \leq \mathrm{mod}\, U_2$$
$$\sigma_1 \leq \mathrm{mod}\, U_2 + \mathrm{mod}\, W_2 .$$

Then,

- $\mathrm{mod}\, U_1 + \mathrm{mod}\, W_1 \geq \frac{1}{2}(\sigma_1 + \sigma_2)$,

- *for every* $\delta > 0$ *there is an* $\epsilon > 0$ *so that if* $\sigma_1 - \sigma_2 \geq \delta$ *and* $\mathrm{mod}\, U_2 \geq \delta$, *then*

$$\mathrm{mod}\,(U_1 \oplus W_1) \geq \frac{\sigma_1 + \sigma_2}{2} + \epsilon .$$

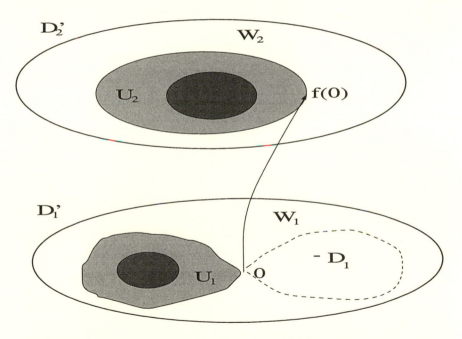

Figure 4.1: The setting of Lemma 4.1.1. Similarly colored regions correspond by f. The critical point 0 is on the boundary of U_1. The disk D_1 is the union of the two shaded areas on the lower level. Notice the twin preimage of D_2 denoted by $-D_1$.

The proof of Lemma 4.1.1 can be found in this monograph in Section 4.2.1. Actually, the first claim is quite easy and just a corollary from the properties we already stated, but the second one is much deeper and will be derived as a consequence of the Modulsatz.

Separation bounds.   Let ϕ be a type I holomorphic box mapping. As usual, B is the domain of the central branch of ϕ and B' is the range of the central branch. Let b be a univalent branch of φ. It is assumed that ϕ was obtained by filling-in from some type II holomorphic box map. In particular, every univalent branch of φ has an analytic continuation as a univalent map onto B'. This continuation will be denoted by $\mathcal{E}_b$. The domain of $\mathcal{E}_b$ is contained in $B' \setminus B$ provided that the domain of b is contained in $B' \setminus B$.

Separating annuli.   The *separating annuli* for b are any five annuli $A_i(b)$, $i = 1, \cdots, 4$, and $A'(b)$, either open or degenerated to Jordan curves, which satisfy the conditions listed below.

- All annuli are contained in B'.

- The complement of $A_2(b)$ contains B in its bounded component and the domain of $\mathcal{E}_b$ in its unbounded component.

- Given $A_2(b)$, $A_1(b)$ is defined as the intersection of B' and the unbounded connected component of the complement of $\overline{A_2(b)}$.

- The annulus $A'(b)$ is uniquely determined as the interior of the set-theoretical difference between the domains of $\mathcal{E}_b$ and b.

- The complement of $A_3(b)$ contains $A'(b)$ in its bounded component and B in its unbounded component.

- Given $A_3(b)$, $A_4(b)$ is defined as the intersection of B' with the unbounded connected component of the complement of $\overline{A_3(b)}$.

Separation symbols.   Remain in the same set-up, i.e. assume that ϕ is a type I holomorphic box mapping derived by filling-in from a type II holomorphic box mapping and that b is a univalent branch of ϕ.

Definition 4.1.1 *A separation symbol $s(D)$ for b is a choice of separating annuli as described above together with a quadruple of numbers $s_i(b)$ for $i = 1, \cdots, 4$ so that the following inequalities hold:*

$$s_2(b) \leq \operatorname{mod} A_2(b)$$

$$s_1(b) \leq \operatorname{mod} A_2(b) + \operatorname{mod} A_1(b)$$

$$s_3(b) \leq \operatorname{mod} A'(b) + \operatorname{mod} A_3(b)$$

$$s_4(b) \leq \operatorname{mod} A'(b) + \operatorname{mod} A_3(b) + \operatorname{mod} A_4(b) \ .$$

Normalized symbols. We will now impose certain algebraic relations among various components of a separation symbol. Choose a number β, set $\alpha := \beta/2$, and pick out $\lambda_1(b)$ and $\lambda_2(b)$. Make sure that

$$\alpha \geq \lambda_1(b), \lambda_2(b) \geq -\alpha$$

and

$$\lambda_1(b) + \lambda_2(b) \geq 0 \ .$$

If these quantities are connected with a separation symbol $s(b)$ as follows

$$s_1(b) = \alpha + \lambda_1(b) \ ,$$

$$s_2(b) = \alpha - \lambda_2(b) \ ,$$

$$s_3(b) = \beta - \lambda_1(b) \ ,$$

$$s_4(b) = \beta + \lambda_2(b) \ ,$$

then we will say that $s(b)$ is normalized with norm β and corrections $\lambda_1(b)$ and $\lambda_2(b)$.

This leads to a definition:

Definition 4.1.2 *For a type I holomorphic box mapping derived by filling-in, a positive number β is called its* separation index *provided that normalized separation symbols with norm β exist for all univalent branches.*

4.1.4   Non-close returns

We will begin the task of constructing and computing separation symbols as box mappings are derived from each other in type I inducing steps.

The set-up.   We assume that a type I holomorphic box mapping ϕ is given which arose by filling-in from a type II map. Let ψ denote the central branch of ϕ with domain B and range B'. Making assumptions about normalized separation symbols for branches of ϕ, we will proceed to construct separation symbols for branches of the mapping $\tilde{\phi}$ obtained from ϕ by a simple inducing step. Our estimates will depend on the combinatorial situation as follows.

We fix a univalent branch b of $\tilde{\phi}$. Let b_0 be the parent branch of b so that $b_0 = p \circ \psi$ where p is a univalent branch of ϕ. Note that p is undefined if the parent branch of b is immediate, see Figure 2.

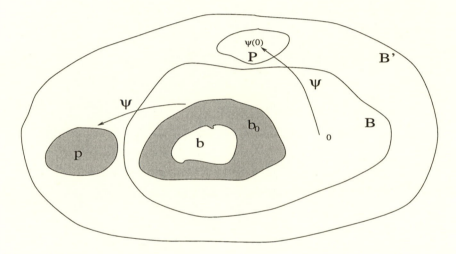

Figure 4.2: The set-up for the construction of separation symbols. For the mapping $\tilde{\phi}$, the box $\tilde{B}'$ is the same as B and $\tilde{B}$ is a smaller domain around 0, not shown in the picture. The domain of the dynamical extension $\mathcal{E}_b$ is contained in the domain of b_0.

Let us define another important branch.

Definition 4.1.3 *Let ϕ be a type I holomorphic box mapping with escaping time $E \geq 1$. We define the* post-critical *branch of ϕ, generically denoted with P, as the (necessarily univalent) branch the domain of which contains $\phi^E(0)$.*

The process of building the separation symbol for b will depend on the combinatorial situation.

Combinatorial analysis. We call a maximal branch b of a type I box mapping ϕ *immediate* provided its parent branch b_0 is immediate. This is consistent with the terminology used to describe the combinatorics of critical filling.

We will distinguish the following combinatorial situations and handle them one by one.

- b is not a maximal branch,

- ϕ makes a close return,

- ϕ makes a non-close return which is subdivided further:

  1. b_0 is immediate,

  2. p and P are independent,

  3. p is subordinate to P,

  4. P is subordinate to p.

Now, let us state the first step in the way proving the growth of separation indices:

Proposition 4 *Let ϕ be a type I holomorphic box mapping with separation index β. Suppose that $\tilde{\phi}$ is derived from ϕ by a type I inducing step showing a non-close return. Assume that ϕ itself was derived from a type II mapping by filling-in, Choose $L \geq 0$ so that the domain of each parent branch of ϕ is separated from the complement of B' by an annulus (possibly degenerate) with modulus L. Suppose that the post-critical branch P of ϕ has a separation symbol $(s_1(P), s_2(P), s_3(P), s_4(P))$ which is normalized with norm β. Suppose further that the domain of P is separated from the complement of B' by an annulus with modulus at least $s_4(P) + \epsilon$.*

Then,

1. *$\tilde{\phi}$ has a separation index β;*

2. *for every L and ϵ positive, there is an $\epsilon_1 > 0$ so that $\beta + \epsilon_1$ is also a separation index for $\tilde{\phi}$;*

3. *$\mathrm{mod}\,(\tilde{B}' \setminus \tilde{B}) \geq \frac{\beta}{4}$.*

We will conclude Proposition 4 after proving a few simple lemmas which involve direct calculations of separation symbols. All lemmas we state in this section have implicit assumptions spelled out above under the heading of the "set-up". We start with general remarks about arithmetics of separation symbols.

Remarks on separation symbols.

Lemma 4.1.2 *Consider a normalized separation symbol*

$$s = (s_1 = \beta/2 + \lambda_1, s_2 = \beta/2 - \lambda_2, s_3 = \beta - \lambda_1, s_4 = \beta + \lambda_2)$$

and suppose that $(s_1, s_2, s_3 + \epsilon, s_4 + \epsilon)$ is another valid separation symbol where $\epsilon > 0$. Than there is a normalized separation symbol with norm $\beta + \frac{\epsilon}{2}$.

Proof:
If we take $\beta' = \beta + \frac{\epsilon}{2}$, $\lambda_1' = \lambda_1 - \frac{\epsilon}{4}$, $\lambda_2' = \lambda_2 + \frac{\epsilon}{4}$, then the normalized separation symbol with norm β' and corrections λ_1' and λ_2' is

$$(s_1, s_2, s_3 + \frac{3}{4}\epsilon, s_4 + \frac{3}{4}\epsilon)$$

which is obviously valid.

$\square$

Lemma 4.1.3 *Suppose that for some univalent branch of a holomorphic type I box mapping one has a normalized separation symbol s with norm $\beta + \epsilon$, $\beta, \epsilon > 0$. Then, for the same branch a normalized separation symbol (s_1', s_2', s_3', s_4') exists with norm β so that $\mathrm{mod}\, A_3(b) + \mathrm{mod}\, A'(b) - s_3' \geq \epsilon/2$.*

Proof:
Consider the normalized symbol s' with norm β and corrections $\lambda_1' = \lambda_1 \frac{\beta}{\beta+\epsilon}$ and $\lambda_2' = \lambda_2 \frac{\beta}{\beta+\epsilon}$ where λ_i are corrections for the symbol s with norm $\beta + \epsilon$. Observe that this will give an algebraically admissible separation symbol, moreover $s_i' \leq s_i$ for $i = 1, 2, 3, 4$. In particular,

$$s_3' = \beta - \lambda_1' = \frac{\beta}{\beta+\epsilon} s_3 .$$

Hence,

$$s_3 - s_3' = s_3 \frac{\epsilon}{\beta + \epsilon} \geq \frac{\epsilon}{2}$$

as $s_3 \geq \frac{\beta + \epsilon}{2}$.

$\square$

b not maximal. The next lemma will allow us to consider separation symbols for maximal branches only.

Lemma 4.1.4 *Suppose that b' is a maximal branch of ϕ and b is subordinate to b'. Assume that the domains of all parent branches of ϕ are separated from the complement of B' by ring domains with modulus at least L. If a separation symbol for b' has bounds (s_1, s_2, s_3, s_4) and a separating annulus $A_3(b')$, then there is a separation symbol for b with bounds*

$$(s_1, s_2, s_3 + L, s_4 + L)$$

with $A_3(b)$ disjoint from the unbounded component of the complement of $A_3(b')$.

Proof:
We can take $A_1(b) = A_1(b')$ and $A_2(b) = A_2(b')$. Likewise, we can certainly adopt $A_4(b) = A_4(b')$. The annulus $A'(b')$ is the preimage of the annulus $B' \setminus B$ by the parent branch of b. The annulus $A'(b)$ is the preimage of the same annulus by the canonical extension of b, so it has the same modulus. By assumption, there is an annulus A of modulus L surrounding $A'(b)$ inside $A'(b')$, hence inside the bounded component of the complement of $A_3(b')$. Thus, we can adopt $A_3(b) = A \oplus A_3(b')$ and the claim of the Lemma follows directly.

$\square$

b_0 immediate.

Lemma 4.1.5 *Suppose that ϕ makes a non-close return and let P have a normalized separation symbol (s_1, s_2, s_3, s_4) with norm β and corrections $\lambda_1(P)$ and $\lambda_2(P)$. We make three claims.*

- *An immediate branch b of $\tilde{\phi}$ has a normalized separation symbol $s(b)$ with norm β and corrections*

$$\lambda_1(b) = \frac{\lambda_2(B)}{2} \, , \; \lambda_2(b) = \frac{\lambda_1(B)}{2} \, .$$

- *If D_P is the domain of P and*

$$\mathrm{mod}\,(B' \setminus D_P) \geq s_4 + \epsilon \, ,$$

then a normalized separation symbol exists for b with norm $\beta + \frac{\epsilon}{4}$.

- *If*

$$\mathrm{mod}\,(B \setminus \tilde{B} \oplus A_2(P)) \geq \mathrm{mod}\,(B \setminus \tilde{B}) + \mathrm{mod}\,A_2(P) + \eta \, ,$$

then the domain of b is surrounded inside $\tilde{B}' = B$ by a ring domain with modulus $s_4(b) + \eta$.

Proof:
The annulus $A_2(b)$ will be the preimage of $A'(P) \oplus A_3(P)$ by the ψ. Then, $A_1(b)$ is the preimage of $A_4(P)$ by ψ. It follows that we can take

$$s_1(b) = \frac{\beta + \lambda_2(P)}{2} \quad \text{and}$$

$$s_2(b) = \frac{\beta - \lambda_1(P)}{2} \, .$$

The annulus $A'(b)$ is conformally equivalent to $B \setminus \tilde{B}$. Hence, if the assumption of the second claim is satisfied, its modulus is at least $s_1(b) + \frac{\epsilon}{2}$. Make $A_3(b)$ equivalent to $A_2(P)$ by the same inverse branch of ψ. We get

$$s_3(b) = \frac{\beta + \lambda_2(P)}{2} + \alpha - \lambda_2(P) \quad \text{and}$$

$$s_4(b) = s_3(b) + \frac{\lambda_1(P) + \lambda_2(P)}{2} = \frac{\beta}{2} + \alpha + \frac{\lambda_1(P)}{2} \, .$$

If the hypothesis of the third claim holds, then

$$\mathrm{mod}\,(A'(b) \oplus A_3(b)) \geq s_3(b) + \eta$$

and so $\mod (A'(b) \oplus A_4(b)) \geq s_4(b) + \eta$ as claimed. If we put

$$\lambda_1(b) = \frac{\lambda_2(P)}{2} \, , \ \lambda_2(b) = \frac{\lambda_1(P)}{2}$$

we get a normalized separation symbol.

If the assumption of the second claim is satisfied, then $s_3(b)$ and $s_4(b)$ can be increased by $\epsilon/2$, hence by Lemma 4.1.2 a normalized symbol can be built with norm $\beta + \frac{\epsilon}{4}$.

$\square$

p subordinate to P.

Lemma 4.1.6 *Suppose that ϕ makes a non-close return, b is a maximal branch of $\tilde{\phi}$, p and P are chosen as explained in the introduction to this section, and p is subordinate to P. Assume that β is a separation index for ϕ and that the domain of the canonical extension $\mathcal{E}_P$ of P is separated from the complement of B' by a ring domain with modulus at least L. Then, a normalized separation symbol exists for b with norm $\beta + \frac{L}{4}$.*

Proof:
The canonical extension $\mathcal{E}_P$ maps the domain of P onto B and the domain of p goes onto the domain of some univalent branch p' of ϕ. Denote $G = \mathcal{E}_P \circ \psi$ and let A be the annulus separating the domain of $\mathcal{E}_P$ from the complement of B'. Set

$$
\begin{aligned}
A_2(b) &= G^{-1}(A_2(p')) \\
A_1(b) &= G^{-1}(A_1(p')) \oplus \psi^{-1}(A) \\
A_3(b) &= G^{-1}(A_3(p') \oplus A'(p')) \\
A_4(b) &= G^{-1}(A_4(p')) \, .
\end{aligned}
$$

The separation symbol is:

$$
\begin{aligned}
\tilde{s}_1 &= \frac{\alpha + \lambda_1(p')}{2} + \frac{L}{2} \\
\tilde{s}_2 &= \frac{\alpha - \lambda_2(p')}{2} \\
\tilde{s}_3 &= \beta + \frac{\alpha - \lambda_1(p') + L}{2} \\
\tilde{s}_4 &= \beta + \frac{\alpha + \lambda_2(p') + L}{2} \, .
\end{aligned}
$$

One checks easily that when $L = 0$ these estimates yield a normalized separation symbol with norm β. By Lemma 4.1.2, the presence of these extra terms allows to construct a normalized symbol with norm $\beta + \frac{L}{4}$.

$\square$

P subordinate to p.

Lemma 4.1.7 *Suppose that ϕ makes a non-close return, b is a maximal branch of $\tilde{\phi}$, p and P are chosen as explained in the introduction to this section, and P is subordinate to p. Assume that β is a separation index for ϕ and that the domain of the canonical extension $\mathcal{E}_p$ of p is separated from the complement of B' by a ring domain with modulus at least L. Then, a normalized separation symbol exists for b with norm $\beta + \frac{L}{4}$.*

Proof:

The canonical extension $\mathcal{E}_p$ maps the domain of p onto B and the domain of P goes onto the domain of some univalent branch P' of ϕ. Denote $G = \mathcal{E}_p \circ \psi$ and let A be the annulus separating the domain of $\mathcal{E}_p$ from the complement of B'.

This case is similar to the case of immediate branches covered by Lemma 4.1.5. If we construct the separating annuli as preimages of the appropriate separating annuli for P' by G, we get the same separation symbol as in the proof of Lemma 4.1.5. The existence of A gives us an extra contribution of $L/2$ in $\tilde{s}_3$ and $\tilde{s}_4$. The proof is concluded by invoking Lemma 4.1.2.

$\square$

P and p are independent.

Lemma 4.1.8 *Suppose that ϕ makes a non-close return and let β be a separation index for ϕ. More specifically, assume that $s(P)$ is a normalized separation symbol for P with norm β, but $\mathrm{mod}\,(A'(P) \oplus A_3(P) \oplus A_4(P)) \geq s_4(P) + \epsilon$ for some $\epsilon \geq 0$. Suppose that the domain of the canonical extension of p is separated from the complement of B' by a ring domain (possibly degenerate) with modulus at least L. We make two claims:*

- *If b is a maximal univalent branch of $\tilde{\phi}$, p and P are chosen as explained in the set-up for this section, and p and P turn out to be independent, then a normalized separation symbol exists for b with norm $\beta + \frac{\epsilon}{4}$.*

- *for every $\alpha_0 > 0$ so that $\mathrm{mod}\,(B' \setminus B) \geq \alpha_0$ and every $L > 0$, there is an $\bar{\epsilon} > 0$ so that the domain of b is separated from the complement of $\tilde{B}'$ by an annulus with modulus at least $\beta + \lambda_2(b) + \bar{\epsilon}$,*

Proof:

Define
$$\lambda := \sup\{\lambda_1(b') : b' \text{ ranges over univalent branches of } \phi\}$$

and $\delta = \max(\lambda_2(P), \lambda - \alpha)$. Since p and P are independent, the domains of their canonical extensions are disjoint. To pick $A_2(b)$, we take $\psi^{-1}(A'(P))$. The annulus $A'(P)$ is biholomorphically equivalent to $B' \setminus B$, and hence its modulus is at least $\alpha + \lambda$. Hence,

$$\tilde{s}_2 = \frac{\alpha + \lambda}{2} \ .$$

Set $A_1(b) = \psi^{-1}(A_3(P) \oplus A_4(P))$. This leads to

$$\mathrm{mod}\,A_1(b) + \mathrm{mod}\,A_2(b) \geq \frac{\beta + \lambda_2(P)}{2} \ . \tag{4.1}$$

We set
$$\tilde{s}_1 := \frac{\beta + \delta}{2} \ .$$

If $\delta = \lambda - \alpha$ this reduces to $\tilde{s}_1 = \tilde{s}_2$, otherwise $\tilde{s}_1 \leq \mathrm{mod}\,A_1(b) + \mathrm{mod}\,A_2(b)$ by estimate (4.1).

As always, $A'(b)$ is determined with modulus at least $\tilde{s}_1$ and if $\epsilon > 0$, then modulus of $A'(b)$ is not less than $\tilde{s}_1 + \frac{\epsilon}{2}$. The annulus $A_3(b)$ will be obtained as the preimage by ψ of $A'(p)$. This has modulus at least $\alpha + \lambda$ in all cases as argued above. The annulus $A_4(b)$ is chosen so that its image is $A_3(p) \oplus A_4(p)$. By induction,

$$
\begin{aligned}
\tilde{s}_3 &= \tilde{s}_1 + \alpha + \lambda = \frac{\beta + \delta}{2} + \alpha + \lambda \\
\tilde{s}_4 &= \tilde{s}_1 + \frac{\alpha + \lambda}{2} + \frac{\beta + \lambda(p)}{2} = \tilde{s}_3 + \frac{\beta + \lambda_2(p) - \alpha - \lambda}{2} \\
&= \beta + \frac{\alpha + \lambda + \lambda_2(p) + \delta}{2} \ .
\end{aligned}
$$

We put $\lambda_1(b) = \frac{\delta}{2}$ and $\lambda_2(b) = \frac{\alpha-\lambda}{2}$. We check that

$$\tilde{s}_3 + \lambda_1(b) = \frac{\beta}{2} + \alpha + \delta + \lambda = \beta + \delta + \lambda \, .$$

Note that

$$\lambda + \delta \geq \lambda + \lambda_2(P) \geq 0 \, .$$

In a similar way one verifies that

$$\tilde{s}_4 - \lambda_2(b) \geq \beta + \frac{\lambda_2(p) + \delta}{2} + \lambda \geq \beta \, .$$

Also, it is clear that $|\lambda_1(b)|, |\lambda_2(b)| \leq \alpha$. Finally,

$$\lambda_1(b) + \lambda_2(b) = \frac{1}{2}(\alpha + \delta - \lambda) \geq \frac{\alpha - \lambda + \lambda - \alpha}{2} = 0.$$

Hence, by possibly decreasing $\tilde{s}_3$ and $\tilde{s}_4$ we get a normalized separation symbol with norm β. Moreover, if $\epsilon > 0$, then $\tilde{s}_3 = \beta - \lambda_1(b) + \epsilon/2$ and $\tilde{s}_4 = \beta + \lambda_2(b) + \epsilon/2$. By Lemma 4.1.2, one can algebraically construct a normalized symbol with norm $\beta + \frac{\epsilon}{4}$ as well. This proves the first claim.

To demonstrate the second claim, apply Lemma 4.1.1 with $U_2 = (D_p \setminus \psi(D_b)) \oplus A'(p)$, $W_2 = A_3(p) \oplus A_4(p)$, and f equal to the central branch of ϕ. We may set

$$\sigma_2 = \frac{\beta + \delta}{2} + \max(\alpha_0, \, \alpha + \lambda)$$

$$\sigma_1 = \max\left(\sigma_2 + L, \, \beta + \lambda_2(p) + \frac{\beta + \delta}{2}\right) \, .$$

Clearly, $\sigma_2 \geq \alpha_0$ and $\sigma_1 - \sigma_2 \geq L$. Hence, by the second claim of Lemma 4.1.1

$$\mathrm{mod}\,(B \setminus D_b) \geq \frac{\sigma_1 + \sigma_2}{2} + \bar{\epsilon} \geq \tilde{s}_4 + \bar{\epsilon},$$

where $\bar{\epsilon}$ depends on L and α_0 only.

$\square$

Proof of Proposition 4. The claim of this Proposition, regarding the separation index β for $\tilde{\phi}$ has already been proved by a separate analysis of all the cases listed at the beginning of this section.

Fix some branch b of $\tilde{\phi}$. It is enough to prove the second claim when b is maximal, according to Lemma 4.1.4. By reviewing the Lemmas 4.1.5, 4.1.6, 4.1.7 and 4.1.8 applied to ϕ, we conclude that all possible maximal branches b of $\tilde{\phi}$ have separation indices increased by ϵ_1 depending only on ϵ and L. So the second claim of Proposition 4 is proved.

To prove the last claim, note that at any rate $s_4(P) \geq \frac{\beta}{2}$. Since $\tilde{B}' \setminus \tilde{B}$ is a 1-to-2 preimage of $A'(P) \oplus A_3(P) \oplus A_4(P)$, the last claim is clear.

4.1.5   Close returns

Construction of separation symbols. Close returns present a difficulty from the combinatorial point of view. We will first show an inductive process of constructing separating annuli and separation symbols corresponding to a simple inducing step with a close return. Later, we will return to this description a number of times as needed. So let us assume that ϕ is a type I holomorphic box mapping and that each univalent branch p of ϕ comes with a separation symbol $s(p)$. Suppose that ϕ shows a close return and look at $\tilde{\phi}$ which is derived from ϕ by a *simple* inducing step. Let ψ denote the central branch of ϕ. Choose a branch b of $\tilde{\phi}$ and let b_0 be the parent branch of b. Then fix p so that $b_0 = p \circ \psi$.

We proceed to define $A_2(b)$ as $\psi^{-1}(A_2(p))$, this determines $A_1(b)$ as well. Then define $A_3(b)$ as an annulus which surrounds the domain of b, is mapped univalently by ψ and

$$\psi(A_3(b)) = A'(p) \oplus A_3(p) . \tag{4.2}$$

Annuli $A'(b)$ and $A_4(b)$ are uniquely determined. Now for the separation symbols, we take

$$s_1(b) = \frac{1}{2}s_1(p) \tag{4.3}$$

$$s_2(b) = \frac{1}{2}s_2(p) \tag{4.4}$$

$$s_3(b) = \frac{1}{2}s_1(p) + s_3(p) \tag{4.5}$$

$$s_4(b) \;=\; \frac{1}{2}(s_1(p) + s_4(p) + s_3(p)) \,. \tag{4.6}$$

All can be easily justified using the first claim of Lemma 4.1.1. If $s(p)$ was a normalized separation symbol with norm β and corrections $\lambda_1(p)$, $\lambda_2(p)$, we can rewrite these formulas as

$$s_1(b) \;=\; \frac{1}{2}(\alpha + \lambda_1(p)) \tag{4.7}$$

$$s_2(b) \;=\; \frac{1}{2}(\alpha - \lambda_2(p)) \tag{4.8}$$

$$s_3(b) \;=\; \beta + \frac{1}{2}(\alpha - \lambda_1(p)) \tag{4.9}$$

$$s_4(b) \;=\; \beta + \frac{1}{2}(\alpha + \lambda_2(p)) \,. \tag{4.10}$$

One immediately observes that formulas (4.7-4.10) determine a normalized separation symbol with norm β and corrections $\lambda_1(b) = \frac{1}{2}(\lambda_1(p) - \alpha))$ and $\lambda_2(b) = \frac{1}{2}(\alpha + \lambda_2(p))$. These observations are summarized as follows.

Lemma 4.1.9 *Let ϕ with the central branch ψ be a type I box mapping which was obtained by filling-in from a type II map. Suppose that ϕ makes a close return that results in the type I box mapping $\tilde{\phi}$. Suppose that b is a univalent branch of $\tilde{\phi}$ and its parent branch is $b_0 = p \circ \psi$ where p is a univalent branch of ϕ. If p has a normalized separation symbol $s(p)$ with norm β, then b has a normalized separation symbol $s(b)$ with the same norm. Moreover, if*

$$\mathrm{mod}\,(A'(p) \oplus A_3(p)) \geq s_3(p) + \epsilon$$

for some $\epsilon > 0$, then $(s_1(b), s_2(b), s_3(b) + \epsilon, s_4(b) + \epsilon)$ is a valid separation symbol for b.

Proof:
The separation symbol $s(b)$ was constructed in the discussion preceding Lemma 4.1.9. Recall equation (4.2) defining $A_3(b)$. It follows that if $\mathrm{mod}\,(A'(p) \oplus A_3(p)) \geq s_3(p) + \epsilon$, then $s_3(b)$ and hence $s_4(b)$ can be increased by ϵ in formulas (4.3-4.6).

$\square$

Hence we get the following.

Proposition 5 *Suppose that ϕ is a type I box mapping with central branch ψ and a separation index β. Let $\tilde{\phi}$ be obtained from ϕ by a type I inducing step. Then $\tilde{\phi}$ also has a separation index β. Moreover, if ϕ makes a close return, B and B' are the domain and range of ψ, respectively, and*

$$\mathrm{mod}\,(B' \setminus \psi^{-1}(B)) \geq \frac{3}{2}\mathrm{mod}\,(B' \setminus B) + \epsilon$$

with $\epsilon > 0$, then $\tilde{\phi}$ has a separation index $\beta + \epsilon_1$ with $\epsilon > 0$ depending on ϵ only.

In view of Proposition 4, it is enough to consider the case when ϕ makes a close return. Let $\phi_0 := \phi, \cdots, \phi_j$ be obtained in a sequence of simple inducing steps with close returns and finally ϕ_j shows a non-close return and yields $\tilde{\phi}$. If Lemma 4.1.9 is used inductively, we see that ϕ_j still has separation index β and the claim is implied by Proposition 4. Finally, if b is a univalent branch of ϕ_1, then $A'(b)$ is conformally equivalent to $B \setminus \psi^{-1}(B)$ and $A_3(b)$ is conformally equivalent to $A'(p) \oplus A_3(p)$ where p is a univalent branch of ϕ. Hence, if

$$\mathrm{mod}\,(B' \setminus \psi^{-1}(B)) \geq \frac{3}{2}\mathrm{mod}\,(B' \setminus B) + \epsilon\,,$$

then $\mathrm{mod}\,(A'(b) \oplus A_3(b)) \geq s_3(b) + \epsilon$ and also $\mathrm{mod}\,(A'(b) \oplus A_3(b) \oplus A_4(b)) \geq s_4(b) + \epsilon$. If $j = 1$, the claim follows directly from Proposition 4. Otherwise, Lemma 4.1.9 shows the situation to be preserved by the sequence of close returns leading to ϕ_j. Proposition 5 is proved.

4.2   Conformal Roughness

One reason for the separation index to increase is given by the second claim of Lemma 4.1.1 in conjunction with Proposition 4. But we need another premise. It turns out that the inequality $\mathrm{mod}\,(C_1 \oplus C_2) \geq \mathrm{mod}\,C_1 + \mathrm{mod}\,C_2$ becomes sharp based on some properties of the curve separating C_1 from C_2. To study this phenomenon, let us first define it.

Definition of conformal roughness.

Definition 4.2.1 *Let w be a Jordan curve in the plane. We say that w is (M, ϵ)-rough if for every pair of open annuli C_1 and C_2 contained in the plane and subject to the conditions*

- *C_1 is contained in the bounded component of the complement of w,*

- *w is contained in the bounded component of the complement of C_2,*

- *the moduli of both annuli are at least M,*

the inequality

$$\mathrm{mod}\,(C_1 \oplus C_2) > \mathrm{mod}\,C_1 + \mathrm{mod}\,C_2 + \epsilon$$

holds.

The content of this definition remains the same if annuli C_1 and C_2 are required to have modulus exactly M instead of at least M.

Teichmüller's Modulsatz. An important analytic tool which can be used to ascertain the roughness of curves is Teichmüller's Modulsatz.

Fact 4.2.1 *Let A_1 and A_2 be two disjoint open annuli situated so that A_1 separates 0 from A_2 while A_2 separates A_1 from ∞. Assume further that both are contained in the ring $A = \{z : r < |z| < R\}$ for some $0 < r < R$. By C denote the set (annulus) of all points from $A \setminus (A_1 \cup A_2)$ separated from 0 and ∞ by $A_1 \cup A_2$. Then, for every $\delta > 0$ there is a number $\epsilon > 0$ with the following property: if*

$$\mathrm{mod}\,A_1 + \mathrm{mod}\,A_2 \geq \mathrm{mod}\,A - \epsilon\,,$$

then a ρ exists for which the ring

$$\{z : \rho < |z| < (1 + \delta)\rho\}$$

contains C.

Fact 4.2.1 follows directly from the "Modulsatz" of [40].

For example, a consequence of Teichmüller's Modulsatz is that every non-analytic Jordan curve is $(M, 0)$-rough for every positive M. This is an example of how the Modulsatz can imply roughness of some curves. In turn, roughness will help us get stronger estimates for separating annuli.

The module theorem. Another classical tool which we use is Teichmüller's module theorem.

Fact 4.2.2 *If the ring domain G separates the points 0 and z_1 from z_2 and ∞, then the bound*

$$\operatorname{mod} G < \log \frac{|z_1| + |z_2|}{|z_1|} + C(|\frac{z_2}{z_1}|)$$

holds where the function C is bounded by $2\log 4$ and goes to 0 as its argument decreases to 0.

Proof:
A version of the theorem which easily implies Fact 4.2.2 is stated on page 56 in [25].

□

4.2.1 Lack of roughness as regularity

We study basic properties of conformal roughness. The leading motive is that a curve which is not rough is in some sense "tame". The tools of this section include the normality properties of families of K-quasiconformal mappings and both theorems of Teichmüller.

Quasi-circles. The first Lemma deals with curves that are not $(M, 0)$-rough. Its main claim says that if w is not $(M, 0)$-rough, then w is Q-quasiconformal where Q is determined by M.

Lemma 4.2.1 *Let A be an annulus. Let G be a canonical univalent mapping from a ring domain bounded by two circles concentric at 0 onto A. Let w be the image by G of some circle centered at 0 so that both components of the complement of w in A are annuli with moduli* *at least $\delta > 0$. For every $\delta > 0$ there is a Q so that G restricted to $G^{-1}(w)$ can be continued to a Q-quasiconformal homeomorphism of the plane. In particular, w is a Q-quasiconformal Jordan curve.*

Proof:
Lemma 4.2.1 is a direct corollary from the following.

Fact 4.2.3 *Let $w_0 : T \to T'$ be a K-quasiconformal mapping and F a compact subset of the domain T. There exists a quasiconformal mapping of the whole plane which coincides with w_0 on F and whose maximal dilatation is bounded by a number depending only on K, T and F.*

Fact 4.2.3 is a verbatim quotation of Theorem 8.1, page 96, from [25].

Normalize our situation so that G^{-1} of w is the unit circle. Then w_0 is defined at least on the ring

$$\{z \ : \ e^{-\delta} < |z| < e^{\delta}\}$$

and we can take this ring as T in Fact 4.2.3. Then w_0 is G restricted to this ring, which is 1-quasiconformal, while F is the unit circle. The claim of Lemma 4.2.1 follows directly.

$\square$

Equicontinuity. Let us recall that if $\mathcal{G}$ is a family of functions from some set $T \subset \hat{\mathbb{C}}$ into the Riemann sphere, then $\mathcal{G}$ is termed *equicontinuous* if for every $z \in T$ and every $\epsilon > 0$ there is a $\delta > 0$ so that $\mathrm{dist}(g(x), g(z)) < \epsilon$ for every $x \in T$ and $g \in \mathcal{G}$ provided that $\mathrm{dist}(z, x) < \delta$, where distances are measured in the spherical metric. Notice that if $\mathcal{G}$ is equicontinuous on T and T is compact, then we can pick the δ independent of z.

Now consider a general fact, see Theorem 4.1 from page 69 in [25]:

Fact 4.2.4 *Let $\mathcal{G}$ be a family of Q-quasiconformal mappings from a domain T into the complex plane. If every mapping $g \in \mathcal{G}$ omits two complex values whose spherical distance is greater than a fixed positive number d (the omitted values need not to be fixed), then $\mathcal{G}$ is equicontinuous in T.*

Limits of curves with decreasing roughness.

Lemma 4.2.2 *Suppose that w_n is a sequence of Jordan curves in the unit disk $D(0,1)$ normalized so that* $\operatorname{diam} w_n = 1$. *Moreover, assume that w_n is not (M, ϵ_n)-rough where ϵ_n tend to zero and M is a fixed positive number. Suppose that w is a compact set such that $w_n \to w$ in the Hausdorff topology. Then there is a univalent mapping G from $T = \{z : |\log|z|| \le M\}$ into $\mathbb{C}$ so that w is an image by G of the unit circle $C(0,1)$. From Lemma 4.2.1, w is a Q-quasiconformal Jordan curve where Q only depends on M.*

Proof:
By the definition of roughness, there exists a sequence of annuli C_n^1 and C_n^2, $\operatorname{mod} C_1^n, \operatorname{mod} C_2^n = M$, so that

$$\operatorname{mod} C_1^n \oplus C_2^n \le \operatorname{mod} C_1^n + \operatorname{mod} C_2^n + \epsilon_n$$

and ϵ_n tends to 0 as n goes to ∞.

Let G_n be a canonical map from the ring T_n into $C_1^n \oplus C_2^n$, where T_n is chosen inversion-symmetric with respect to the unit circle. In particular, $T_n \supset T := \{z : |\log|z|| < M\}$. By Teichmüller's Modulsatz, the curves $G_n^{-1}(w_n)$ tend to the unit circle in the Hausdorff topology. Since every G_n avoids ∞ and another point inside the unit disk, the family G_n is equicontinuous in T (see Fact 4.2.4). Since G_n are uniformly equicontinuous on small neighborhoods of the unit circle, the Hausdorff distance from w_n to $G_n(C(0,1))$ goes to 0. By picking a subsequence we may assume that $G_n \to G$ uniformly on compact subsets of T. To see that G is univalent observe that $\operatorname{diam} w_n = 1$ and the annuli C_2^n all avoid ∞. Then the curve $G_n(C(0, \exp(M/2)))$ is in the distance from the boundary of C_2^n that is bounded away from 0 in terms of M. This follows from Teichmüller's module theorem, see Fact 4.2.2. So, the derivatives of G_n' remain bounded away from 0 on this curve. Then, clearly $w = G(C(0,1))$.

$\square$

Lemma 4.2.3 *Let W be an annulus in the plane. At the same time, assume that the outer component of the boundary of W is a Jordan curve w. Assume that $\operatorname{mod} W \ge \Delta$. Finally, suppose that the distance between the two connected components of the complement of*

W *is at least* $\lambda \cdot \operatorname{diam} W$. *Let* H_w *be the canonical map from the ring* $\{z : -\operatorname{mod} W < \log|z| < 0\}$ *onto* W.

    For every Δ, M, δ *and* λ *positive, there are positive numbers* ϵ_1 *and* ϵ *so that either* w *is* (M, ϵ)-*rough or the Hausdorff distance from* $H_w\left(C(0, e^{-\epsilon_1})\right)$ *to* w *is less than* $\delta \cdot \operatorname{diam} w$.

Proof:

It is enough to prove the lemma when $\operatorname{mod} W = \Delta$. Suppose that the claim of Lemma 4.2.3 is false. Then there exist positive constants M, δ, Δ, λ, a sequence of annuli W_n, and two sequences $\epsilon_n > 0$ and $\epsilon_n' > 0$ which tend to 0 so that

- $\operatorname{diam} W_n = 1$ and $W_n \subset D(0, 1)$,

- $\operatorname{mod} W_n = \Delta$,

- the outer connected component of the boundary of W_n is a Jordan curve w_n not (M, ϵ_n)- rough,

- the distance between the connected components of the complement of W_n is larger than λ for all n.

- the Hausdorff distance from w_n to $H_{w_n}\left(C(0, \exp(-\epsilon_n'))\right)$ is at least δ.

    By choosing a subsequence, we can assume that w_n converge in the Hausdorff topology. If w is the limit of w_n, then Lemma 4.2.2 applies to show that w is the image of the unit circle by a univalent map, at any rate a Jordan curve.

    The canonical maps H_{w_n} also form a normal family. Since the distance between the connected components of the complement is bounded away from 0, the limits are univalent. Without loss of generality, $H_{w_n} \to H$ where H is univalent. The range of H is an annulus with modulus Δ and the outer connected component of its boundary is w. By Caratheodory's theorem H can be extended continuously to the circle $C(0, 1)$. But then for some $\epsilon' > 0$, the Hausdorff distance from w to $H\left(C(0, \exp(-\epsilon'))\right)$ is below $\delta/2$ contrary to our assumption.

□

    We will several times use the following corollary to Lemma 4.2.3.

Corollary 4.1 *Let W satisfy the hypothesis of Lemma 4.2.3 with some Δ and λ, and U be an annulus contained in W in such a way that it separates the components of the boundary of W. Let u be the outer connected component of the boundary of U. Suppose that $\mod U + \epsilon_1 \geq \mod W$. Then for every M, Δ, δ and λ positive, there are $\epsilon, \epsilon_1 > 0$ so that either w is (M, ϵ)-rough or the Hausdorff distance between u and w is less than $\delta \cdot \operatorname{diam} w$.*

To prove Corollary 4.1, consider the canonical map H from the ring $\{z : -\Delta < \log |z| < 0\}$ onto W. By Teichmüller's Modulsatz, applied to the pair of annuli cut from W by u, $H^{-1}(u)$ is contained in a ring whose width goes to 0 with ϵ_1, and whose distance from the unit circle is no more than ϵ_1. So, the Hausdorff distance from $H^{-1}(u)$ to the unit circle goes to 0 with ϵ_1, and Corollary 4.1 follows from Lemma 4.2.3.

Bounded turning and pinched curves. The bounded turning property is formulated as Fact 2.3.2 in Chapter 2. We have the following lemma as a corollary to the bounded turning property.

Lemma 4.2.4 *Let w be Jordan curve in the plane which is invariant with respect to a non-trivial subgroup of the group of rotations about 0.*

For every $M > 0$ there are L and $\epsilon > 0$ so that if $|z_1|/|z_2| \geq L$ for some $z_1, z_2 \in w$, then w is (M, ϵ)-rough.

Proof:
Suppose that the claim of the lemma is not true. After normalization by affine transformations we may assume that there exists a sequence of curves w_n which are not (M, ϵ_n)-rough and have the diameter 1. Now, we apply Lemma 4.2.2 to get as the limit a $Q(M)$-quasiconformal curve w. Of course, w is still invariant by the same subgroup of rotations. But $|z_2|/|z_1|$ is obviously bounded for $z_1, z_2 \in w$ in terms of $Q(M)$, based on Fact 2.3.2. The symmetry of w plays a role because when $|z_2/z_1|$ reaches a maximum, we can find other points $z_1', z_2' \in w$ so that $|z_1'| = |z_1|$, $|z_2'| = |z_2|$ and z_1, z_2, z_1', z_2' are cyclically ordered on w.

$\square$

Proof of Lemma 4.1.1. We now give the proof of Lemma 4.1.1 which has long been postponed.

To prove the first claim, consider the canonical mapping H from the ring $\{z : d < |z| < 1\}$ onto W_2. Recall that 0 is the critical point of f and set $t = |H^{-1}(f(0))|$. This splits W_2 into two annuli, $V_i = H(\{z : d < |z| < t\})$ and $V_o = H(\{z : t < |z| < 1\})$. We have $\operatorname{mod} W_2 = \operatorname{mod} V_i + \operatorname{mod} V_o$. There is an annulus U_i surrounding U_1 which is mapped by f conformally onto V_i and the annulus U_o mapped onto V_o by f as a degree 2 cover. Note that $U_o \oplus U_i = W_1$. So,

$$
\begin{aligned}
\operatorname{mod}\left(U_1 \oplus W_1\right) &\geq \operatorname{mod} U_2 + \operatorname{mod} V_i + \frac{1}{2}\operatorname{mod} V_o \qquad (4.11) \\
&\geq \frac{1}{2}(\operatorname{mod} U_2 + \operatorname{mod} W_2) + \frac{1}{2}\operatorname{mod} U_2 \\
&\geq \frac{1}{2}(\sigma_1 + \sigma_2)
\end{aligned}
$$

which proves the first claim.

For the second claim, look at the curve

$$
w_\eta := f^{-1}\left(H^{-1}(\{z : |z| = t + \eta\})\right)
$$

for a small and positive η. It is a centrally symmetric Jordan curve, but as η decreases to 0, the corresponding w_η develop a pinching point at 0. Comparing this with Lemma 4.2.4, we see that for every $M > 0$ there is an $\eta_0 > 0$ and $\epsilon > 0$ for which w_η is (M, ϵ)-rough for every $\eta < \eta_0$. Make M equal to half the minimum of $\operatorname{mod} U_o = \frac{1}{2}\operatorname{mod} V_o$ and $\operatorname{mod} U_1 \geq \operatorname{mod} U_2$. Now consider two cases.

$\operatorname{mod} V_o \geq \frac{1}{2}(\sigma_1 - \sigma_2)$. In this case, $M \geq \frac{\delta}{8}$ where δ is the parameter from Lemma 4.1.1. It follows that

$$
\operatorname{mod}\left(U_1 \oplus W_1\right) \geq \operatorname{mod} U_1 + \operatorname{mod} U_i + \operatorname{mod} U_o - \eta + \epsilon
$$

where ϵ depends on δ and η can be made arbitrarily small without affecting ϵ. We can follow the calculation (4.11) to get

$$
\operatorname{mod}\left(U_1 \oplus W_1\right) \geq \frac{1}{2}(\sigma_1 + \sigma_2) + \epsilon
$$

so the second claim of Lemma 4.1.1 holds in this case.

Otherwise. We simply compute

$$\mod(U_1 \oplus W_1) \geq \mod U_2 + \mod V_i + \frac{1}{2}\mod V_o$$

$$\geq \sigma_1 - \frac{1}{2}\mod V_o \geq \sigma_1 - \frac{1}{4}(\sigma_1 - \sigma_2)$$

$$= \frac{1}{2}(\sigma_1 + \sigma_2) + \frac{1}{4}(\sigma_1 - \sigma_2)$$

and the claim of Lemma 4.1.1 holds with $\epsilon := \frac{\delta}{4}$. This ends the proof.

4.2.2 Quasi-invariance of roughness

The key notion used in this section is (M, ϵ)-roughness. The issue addressed in this section is how conformal roughness is affected when the curve is mapped by an analytic map. We are interested in knowing what occurs when a curve is lifted to a holomorphic cover of its neighborhood. This is the typical relation which occurs between the boundaries of the domain and range of the central branch of the box mapping is obtained by simple inducing step. Conformal roughness turns out to be quasi-invariant, that is it still holds for the lifted curve, though perhaps with parameters changed in an estimable way. This section is devoted to a proof of this statement.

Lemma 4.2.5 *Suppose that w is a Jordan curve in $\mathbb{C}$. Let w be contained in the closed ring $r \leq |z| \leq 1$. Let Γ be a subgroup of the group of rotations about the origin.*

For every $M, r > 0$ there is an $M' > 0$ and for every M, r and $\epsilon' > 0$ there is an $\epsilon > 0$ so that if w is not (M, ϵ)-rough, then there are nesting annuli A_1 and A_2 in the plane, disjoint, sharing w as a connected component of their boundaries and each fixed by Γ, so that $\mod A_1 = \mod A_2 = M'$ and $\mod(A_1 \oplus A_2) \leq 2M' + \epsilon'$.

Proof:
The whole point of this lemma is making A_1 and A_2 fixed under Γ. If Γ is trivial, annuli C_1 and C_2 which exist by contradicting Definition 4.2.1 will do, with $M' = M$ and $\epsilon = \epsilon'$ in this case. To prove Lemma 4.2.5, it is sufficient to show the following.

Pick M and ϵ'. Suppose that v_n is a sequence of Jordan curves invariant under Γ and contained in the closed ring $r \leq |z| \leq 1$ chosen so that v_n is not (M, ϵ_n)-rough where ϵ_n are all positive and tend to 0 with n. Then for some n nesting annuli A_1 and A_2 exist, fixed by Γ, disjoint and sharing v_n as a connected component of the boundary, both with modulus at least M' which depends only on M, and

$$\mathrm{mod}\,(A_1 \oplus A_2) \leq \mathrm{mod}\,A_1 + \mathrm{mod}\,A_2 + \epsilon' \,.$$

By choosing a subsequence if needed, we can assume that v_n converge in the Hausdorff metric. Then Lemma 4.2.2 tells us that the limit v is a Jordan curve, moreover that it is the image of the unit circle under a univalent map G which is defined on the ring $T := \{z : |\log|z|| < M\}$. Also, v is invariant under Γ and contained in the closed ring $\{z : r \leq |z| \leq 1\}$.

By Teichmüller's module theorem, the range of G contains the set of all z with distance from v less than η, where η only depends on M. Consider the Riemann map $\mathcal{R}$ from the unit disk onto the bounded connected component of the complement of v which fixes 0. Since v is not $(M, 0)$-rough, by taking W equal to the inner component of the complement of v punctured at 0 we are in the position to apply Corollary 4.1. Notice that the canonical map of W is just $\mathcal{R}$ restricted to the punctured disk. We get that for some η_1 which only depends on η and M, $\mathrm{dist}(\mathcal{R}(z), v) < \eta$ whenever $|z| \geq \eta_1$. Let us define annulus

$$A_1' = \mathcal{R}(\{z : \eta_1 < |z| < 1\}) \,.$$

Since $\mathcal{R}$ commutes with Γ, then A_1' is fixed by Γ. Also, its modulus is $-\log \eta_1$, hence only dependent on M.

Now consider the anti-holomorphic map

$$\rho := G \circ R \circ G^{-1}$$

where R is the inversion about the unit circle. Define $A_2' := \rho(A_1')$. Note that $\mathrm{mod}\,A_1' = \mathrm{mod}\,A_2'$ and

$$
\begin{aligned}
\mathrm{mod}\,(A_1' \oplus A_2') &= \mathrm{mod}\,(G^{-1}(A_1') \oplus G^{-1}(A_2')) \\
&= \mathrm{mod}\,G^{-1}(A_1') + \mathrm{mod}\,G^{-1}(A_2') \\
&= 2\mathrm{mod}\,A_1' \,.
\end{aligned}
$$

The middle equality follows since $G^{-1}(A'_1)$ and $G^{-1}(A'_2)$ correspond by inversion R. It is then easy to see that the canonical map from a ring $x < |z| < x^{-1}$ onto $G^{-1}(A'_1) \oplus G^{-1}(A'_2)$ fixes the unit circle. Next, we will show that $\chi(A'_2) = A'_2$ for all $\chi \in \Gamma$. Consider the holomorphic map $\rho \circ \chi \circ \rho$ which is defined at least on some connected neighborhood of v including A'_2. On v, we have $\chi(z) = \rho \circ \chi \circ \rho(z)$, and so these maps are equal on A'_2 by the identity principle of holomorphic maps. Thus

$$\chi(A'_2) = \rho(\chi(\rho(A'_2))) = \rho(\chi(A'_1)) = \rho(A'_1) = A'_2 \, .$$

Let α_1 denote the connected component of the boundary of A'_1 other than v and α_2 the connected component of the boundary of A'_2 other than v. By the continuity of modulus with respect to the Hausdorff topology, for every ϵ' we can find n so that the modulus of the annulus A_1 bounded by α_1 and v_n is between $\operatorname{mod} A'_1 - \frac{\min(\epsilon', \log \eta_1^{-1})}{2}$ and $\operatorname{mod} A'_1 + \frac{\min(\epsilon', \log \eta_1^{-1})}{2}$ and the same holds true for the modulus of the annulus A_2 bounded by v_n and α_2. Both these annuli are bounded by Jordan curves fixed by Γ. Also, $\operatorname{mod} A_1, A_2 \geq \frac{1}{2} \log \eta_1^{-1}$ and

$$\operatorname{mod}(A_1 \oplus A_2) \leq \operatorname{mod} A_1 + \operatorname{mod} A_2 + \epsilon' \, .$$

Our auxiliary statement and hence the Lemma both follow.

$\square$

The next lemma is a consequence of Corollary 4.1.

Lemma 4.2.6 *Suppose that w and α are Jordan curves. Assume that both separate 0 from ∞ and are fixed by some subgroup Γ of the group of rotations about the origin. Let $\operatorname{dist}(0, w) \geq \lambda$. Suppose that α belongs to the bounded connected component of the complement of w so that α and w delimit an annulus A. Let $\operatorname{mod} A \geq \Delta$. For every $M, \delta, \Delta, \lambda > 0$ there are numbers $\epsilon, \epsilon_1 > 0$ so that if w is not (M, ϵ)-rough then A can be represented as $A' \oplus A''$ and the following conditions hold:*

- *A' and A'' are nesting annuli, both invariant by Γ,*

- *$\operatorname{mod} A' + \operatorname{mod} A'' = \operatorname{mod} A$,*

- $\mod A' \geq \epsilon_1$,

- *the Hausdorff distance from A' to w is less than $\delta \cdot \operatorname{diam} w$.*

Proof:
We are in the position of Corollary 4.1 with $W := A$ and parameters $M, \delta, \Delta, \lambda$ playing the same role. Choose ϵ and ϵ_1 according to Corollary 4.1. Let G be the canonical map from the ring $r < |z| < 1$ onto A so that $\lim_{|z| \to 1} G(z) \in w$. Then G commutes with Γ. Define

$$A' = G(\{z : -\epsilon_1 < \log |z| < 0\}) .$$

The claims of Lemma 4.2.6 are readily verified.

$\square$

Lemma 4.2.7 *Choose an integer $\ell > 1$. Let w be a Jordan curve in $\mathbb{C}$, separating 0 from ∞ and invariant under the rotation by $2\pi/\ell$ about the origin. Let U_δ be the set of all z whose distance from w is less than $\delta \cdot \operatorname{diam} w$. Assume that function ψ is defined on U_δ and representable as $\psi(z) = h(z^\ell)$ with h univalent.*

We claim that for every ℓ and $M, \delta, \epsilon' > 0$ there are $M_1, \epsilon > 0$, where M_1 only depends on M and δ, so that if $w' = \psi(w)$ is (M_1, ϵ')-rough, then w is (M, ϵ)-rough.

Proof:
Normalize w so that its Hausdorff distance from the origin is 1. Then $\operatorname{diam} w = 2$. In view of Lemma 4.2.4 we can pick an $r > 0$ depending only on M and assume without loss of generality that w does not intersect $C(0, r)$. Without loss of generality, $\delta < r/2$ so that $0 \notin U_\delta$. Let Γ be the group generated by the rotation by $2\pi/\ell$ about the origin.

With w we are in the situation of Lemma 4.2.5. For every M and $\bar{\epsilon}$ pick M' and ϵ so that if w is not (M, ϵ)-rough, then there are annuli A_1 and A_2, on alternate sides of w and nesting, invariant under Γ, so that $\mod A_1 = \mod A_2 = M'$ and

$$\mod (A_1 \oplus A_2) \leq 2M' + \bar{\epsilon} . \tag{4.12}$$

Moreover, M' is independent of $\bar{\epsilon}$.

To proceed with the proof, we would like A_1 and A_2 to be contained in U_δ. This may require shrinking them. Suppose that A_1 belongs to the bounded connected component of the complement of w. Apply Lemma 4.2.6 with $\Delta := M'$ and M, δ the same as in the ongoing argument. This gives us the splitting $A_1 = A_1' \oplus A_1''$ with $A_1' \subset U_\delta$. In the case of A_2, we have to map the picture by the inversion about the unit circle. The image of w under this inversion has diameter bounded by $2/r$. Applying Lemma 4.2.6 to the transformed situation with $\delta := \delta/r$ and other parameters as before, we get the splitting $A_2 = A_2' \oplus A_2''$ with $A_2' \subset U_\delta$. Looking at other claims of Lemma 4.2.6, we can also see that $\mathrm{mod}\, A_1'', \mathrm{mod}\, A_2'' \geq M''$ where M'' only depends on M and δ from the hypothesis of the current Lemma. Next, we estimate the moduli. By superadditivity,

$$\mathrm{mod}\,(A_1' \oplus A_2') \leq \mathrm{mod}\,(A_1 \oplus A_2) - \mathrm{mod}\, A_1'' - \mathrm{mod}\, A_2'' \,. \qquad (4.13)$$

From inequality (4.12) and the claim of Lemma 4.2.6 about the splitting

$$
\begin{aligned}
\mathrm{mod}\,(A_1 \oplus A_2) &\leq \mathrm{mod}\, A_1 + \mathrm{mod}\, A_2 + \bar{\epsilon} \\
&= \mathrm{mod}\, A_1' + \mathrm{mod}\, A_2' + \mathrm{mod}\, A_1'' + \mathrm{mod}\, A_2'' + \bar{\epsilon} \,.
\end{aligned}
$$

From this and estimate (4.13),

$$\mathrm{mod}\,(A_1' \oplus A_2') \leq \mathrm{mod}\, A_1' + \mathrm{mod}\, A_2' + \bar{\epsilon} \,.$$

But now ψ restricted to $A_1' \oplus A_2'$ is a holomorphic degree ℓ cover which sends w to w', A_1' to C_1' and A_2' to C_2'. Since all moduli are multiplied by ℓ,

$$\mathrm{mod}\,(C_1' \oplus C_2') \leq \mathrm{mod}\, C_1' + \mathrm{mod}\, C_2' + \ell\bar{\epsilon} \,.$$

The number $\bar{\epsilon}$ was arbitrary and can be set to ϵ'/ℓ. If we choose M_1 in Lemma 4.2.7 to be $\ell M''$, we see that w' is not (M_1, ϵ')-rough. This contradiction concludes the proof.

$\square$

4.3   Growth of the Separation Index

4.3.1   Consequences of roughness

As usual, given a box mapping ϕ we denote by ψ its central branch and by B' and B, the range and the domain of ψ, respectively. D_b stands for the domain of a univalent branch b. Since ϕ is symmetric, ψ can be represented as $h(z^2)$ where h is a univalent map onto B'.

Conformal roughness increases separation.

Lemma 4.3.1 *Suppose that ϕ is a type I holomorphic box mapping with separation index $\beta \geq \underline{\beta}$. Let $\mathrm{mod}\,(B' \setminus B) \geq \alpha_0$. For every $\alpha_0, \beta_0, \epsilon$, all positive, there are $M > 0$, independent of ϵ, and $\eta > 0$ so that if the boundary of B is (M, ϵ)-rough, then the map ϕ_4 obtained from ϕ by four type I inducing steps has a separation index $\beta + \eta$.*

Proof:
If ϕ makes a close return and M is chosen to be less than $\alpha_0/2$, then

$$\mathrm{mod}\,(B' \setminus \psi^{-1}(B)) \geq \frac{3}{2}\mathrm{mod}\,(B' \setminus B) + \epsilon \,.$$

Invoking Proposition 5, we see that the separation index increases already after the first step. On the hand, if ϕ makes a non-close return, then the roughness of the boundary of the central domain is quasi-preserved. More precisely, denote by ϕ_i the map obtained from ϕ by i type I inducing steps. Let ψ_i, B_i and B'_i denote the central branch of ϕ_i, its domain and range, respectively. Then ψ_1 maps ∂B_1 onto ∂B but extends as a quadratic-like map onto the range B'. It follows that $\psi_1(z) = h_1(z^2)$ on some domain which contains B_1 encompassed by an annulus with modulus at least $\alpha_0/2$. By Teichmüller's module theorem, this domain contains all points with distance from B_1 less than $\delta \cdot \mathrm{diam}\, B_1$, where δ depends on α_0. We are in the situation of Lemma 4.2.7. As the result, ϕ_1 still satisfies the hypotheses of Lemma 4.3.1. Indeed, by Proposition 4, the parameter α_0 may be taken equal to $\beta_0/4$, the boundary of B_1 is (M', ϵ')-rough and ϵ' depends on the original $\epsilon, \beta_0, \alpha_0$, and finally M' can be set to an arbitrary value by choosing M appropriately. Hence, without loss of generality we may assume that $\phi_1, \cdots, \phi_3$ also make non-close returns and the boundaries of B_i are rough.

Observe that all parent domains of ϕ_1 are separated from the boundary of B'_1 by annuli with modulus at least $\alpha_0/2$. From Lemmas 4.1.6-4.1.8, if b is a univalent branch of ϕ_i, $i = 2, 3$ and not immediate, then b has a normalized separation symbol $s(b)$ and the domain of b is separated from the border of B'_i by an annulus with modulus $s_4(b) + \epsilon_1$ with $\epsilon_1 > 0$ depending only on α_0, β_0. By Proposition 4, we are done unless the critical value $\psi_i(0)$ belongs to one of the immediate branches. Now look at the third claim of Lemma 4.1.5 applied to ϕ_2. Let P be the post-critical branch of ϕ_2, which is immediate. By the inspection of separation symbols constructed in Lemma 4.1.5, we see that P has a normalized separation symbol with norm β and the correction $\lambda_2(P) \leq \beta/4$, Hence, $\mod A_2(P) \geq \beta_0/4$. Now if ∂B_2 is $(\beta_0/4, \epsilon')$-rough, then the hypothesis of the third claim of Lemma 4.1.5 is fulfilled with $\eta := \epsilon'$. As a consequence, also the domains of univalent branches b of ϕ_3 are separated from the boundary of B'_3 by annuli with moduli at least $s_4(b) + \eta$. If the critical value of ϕ_3 again visits one of these immediate branches, the separation index of ϕ_4 grows to $\beta + \eta/4$ by the second claim of Lemma 4.1.5.

This analysis has covered all cases showing that an increase in the separation index is inevitable.

$\square$

Roughness or additional separation. This section contains the final step of proof, showing that in the typical situation an increase in the separation index can only be avoided if the boundary of the central domain is made conformally rough.

In the forthcoming Lemma, observe that we do not assume any positive lower bound on the modulus of $B' \setminus B$.

Lemma 4.3.2 *Let ϕ be a box mapping which makes a non-close return and has a separation index $\beta \geq \beta_0$. Suppose that the representation $\psi(z) = h(z^2)$ with h univalent holds on some domain U which contains all points with the distance from B less than $\Delta \cdot \operatorname{diam} B$. Also, we assume that there exists $L \geq 0$ so that the domain of each parent branch of ϕ is separated from $\partial B'$ by an annulus with modulus L.*

Let $\tilde{\phi}$ be derived from ϕ by one simple inducing step. Then for every L, M, Δ and β_0 positive there exist $\eta_1 > 0$ and $\eta_2 > 0$ so that *the following alternative holds true:* $\tilde{\phi}$ *has*

- *the domain D_P of the postcritical branch P is surrounded inside B' by an annulus with modulus at least $s_3(P) + \eta_1$, or*

- *the boundary of $\tilde{B}'$ is (M, η_2)-rough.*

Proof:
Suppose that that $\operatorname{mod}(B' \setminus P) \le s_3(P) + \nu$, where ν is a small parameter to be specified later. By superadditivity of moduli, this means that

$$\operatorname{mod}(\tilde{B}' \setminus \tilde{B}) \le \operatorname{mod}(\psi^{-1}(A_3(P) \oplus A'(P)) + \frac{1}{2}\nu.$$

We use Corollary 4.1. Put $U := \psi^{-1}(A_3(P) \oplus A'(P))$ and $W = (\tilde{B}' \setminus \tilde{B})$. Observe, that the boundaries of W and U are symmetric with respect to 0 and $\operatorname{mod} W \ge \beta_0/4$. By Corollary 4.1, for every M, δ and β_0 positive there exist η_2 and ϵ_1 so that if

$$\operatorname{mod} U + \epsilon_1 \ge \operatorname{mod} W,$$

then either

- $\partial \tilde{B}'$ is (M, η_2)-rough or

- the Hausdorff distance between the connected components of the boundary of $\psi^{-1}(A_4)$ is less than $\delta \cdot \operatorname{diam} \partial \tilde{B}'$.

We set $\nu := \epsilon_1$. Observe that ν still depends on δ. We will show how to dispense with this dependence by choosing suitable $\delta > 0$ in terms of β_0 and M. To this aim we will show that under some additional conditions on regularity of $\partial B'$, there exists $\delta_1 > 0$ so that for every $\delta \le \delta_1$ the boundary of $\tilde{B}'$ is (M, η_2)-rough. Then it will be enough to adjust $\nu = \epsilon_1$ to $\delta := \delta_1$.

Step I. Set $\delta := \Delta$ in the assumptions of Lemma 4.2.7. Then for every M and Δ there exists M_0 and if one additionally specifies an $\epsilon > 0$, then there exists $\eta_2 > 0$ such that if $\partial B'$ is (M_0, ϵ)-rough then ∂B is (M, η_2)-rough. Here ϵ is a free parameter whose value will be specified in the second step.

Step II. Now we invoke Lemma 4.2.4 for M_0 as a free parameter. This yields K and ϵ_0 so that if $|z_1|/|z_2| > K$ for some z_1 and $z_2 \in \partial B'$, then $\partial B'$ is (M_0, ϵ_0)-rough. We fix the parameter ϵ from the first step to be equal to ϵ_0. Observe that ϵ_0 depends only on M and β_0, and thus η_2 is also a function of only these two parameters. Therefore, Step I and Step II imply the second alternative in the claim of Lemma 4.3.2 unless $\partial B'$ has a fairly round shape.

The choice of δ_1. The only possibility which does not lead immediately to (M, η_2)-roughness of $\partial B = \partial \tilde{B}'$ occurs when for all points $z_1, z_2 \in \partial B'$

$$\frac{|z_1|}{|z_2|} \leq K \,,$$

and K depends solely on M.

Let $\delta_0 \cdot \text{diam}\,\partial B'$ be the Hausdorff distance between the connected components of the boundary of $A_4(P)$. Since $\psi = h(z^2)$, and h is of bounded distortion in terms of β_0, then, by the standard reasoning, for every $\delta_0 > 0$ there exists a positive constant δ_1 so that the Hausdorff distance between the connected components of the boundary of $\psi^{-1}(A_4(P))$ is less than $\delta_1 \cdot \text{diam}\,\partial B$.

Let $\delta_0 < \frac{1}{K}$. The outer connected component of the boundary of $A_4(P)$ is symmetric about 0 and hence for every $z \in A_4(P)$ we have that

$$|z| \geq \left(\frac{1}{2K} - \frac{\delta_0}{2} \right) \cdot \text{diam}\,\partial B' > 0 \,,$$

which leads to the contradiction since $A_4(P)$ does not separate 0 from $\partial B'$.

We set $\delta_0 := \frac{1}{2K}$ and determine the corresponding δ_1 which is a function of K and β_0 only. Therefore, setting $\delta := \delta_1$, we conclude that the boundary of B' is (M, ϵ)-rough provided that $\text{mod}\,(B' \backslash P) \leq s_4 + \nu$ and ν depends only on L, M, Δ and β_0, so it can be used as η_1.

$\square$

Lemma 4.3.3 *Suppose that ϕ_0 is a type I holomorphic box mapping with a separation index $\beta \geq \beta_0$. If B and B' denote the domain and range of the central branch of ϕ, respectively, suppose that $\text{mod}\,(B' \backslash$*

$B) \geq \alpha_0$. *Suppose that ϕ makes a close return which after a type I inducing step results in the box mapping $\tilde{\phi}$.*

For every β_0, α_0, M positive, there are positive numbers δ, ϵ so that one of the following holds:

- *$\tilde{\phi}$ has a separation index $\beta + \delta$,*

- *the boundary of the central domain $\tilde{B}$ of $\tilde{\phi}$ is (M, ϵ)-rough.*

Proof:

Suppose that ϕ shows a close return with escaping time E and φ_i, $i = 0, \ldots E - 1$, are derived from ϕ_0 by i simple inducing steps.

We set $\varphi := \varphi_{E-1}$ and $\hat{\varphi} = \varphi_{E-2}$. Let B'_φ denote the range of φ. Then φ shows a non-close return and parent branches of φ are separated from the boundary of B'_φ by annuli with moduli at least $\alpha_0/2$. This is because every univalent branch of φ has a univalent extension onto B' with the domain contained in B_φ. Let P denote the post-critical branch of φ. Now Lemma 4.3.2 implies that either

- $\operatorname{mod}(B'_{\tilde{\phi}} \setminus P) \geq s_3(P) + \delta'$ or

- $\partial \tilde{B}'$ is (M', ϵ')-rough where M' needs to be specified depending on α_0, β_0 and M.

To finish the proof in the latter case of the alternative, we need to demonstrate that $\partial \tilde{B}$ is rough, not $\partial \tilde{B}'$. This, however, follows by Lemma 4.2.7. Indeed, if $\tilde{\psi}$ is the central branch of $\tilde{\phi}$, then $\tilde{\psi}^{-1}(\tilde{B}') = \tilde{B}$ and $\tilde{\psi}(z) = h(z^2)$ with h univalently mapping onto B'. By Teichmüller's module theorem, δ in Lemma 4.2.7 can be set depending only on α_0. Then for every $M > 0$ we can pick an $M' > 0$ and $\epsilon' > 0$ so that if $\partial \tilde{B}'$ is (M', ϵ')-rough, then $\partial \tilde{B}$ is (M, ϵ)-rough.

Let us now deal with the first case of the alternative. We will prove that the separation index grows for $\tilde{\phi}$. The parent branch of P is of the form $P^* \circ \psi$ where P^* is a univalent branch of $\hat{\varphi}$. Suppose that P^* has a normalized separation symbol with norm β set up so that $s_4(P^*) - s_3(P^*) = \eta \geq 0$.

Consider the separation symbol for P obtained in the standard way described for close returns. Apply Lemma 4.1.1 with $U_1 := A_3(P)$, $W_1 := A_4(P)$ and $f := \psi$. Then $U_2 := A'(P^*) \oplus A_3(P^*)$ and $W_2 := A_4(P^*)$. Choose $\sigma_2 := s_3(P^*)$ and $\sigma_1 := s_4(P^*)$. Since

$\mod U_2 \geq \beta_0/4$, the second claim of Lemma 4.1.1 asserts the existence of $\kappa > 0$ depending on $\beta_0 > 0$ and $\eta > 0$ so that

$$\mod (B'_\varphi \setminus D_P) \geq s_4(P) + \kappa .$$

On the other hand, by the construction of separation symbols for close returns, we get $s_4(P) - s_3(P) = \frac{\eta}{2}$.

Let us now compare η with δ'. If $\eta \geq \delta'$, then κ is bounded away from 0 in terms of δ'. Otherwise, since the first case of the alternative holds, we have that

$$\mod (B_\varphi \setminus D_P) \geq s_3(P) + \delta' \geq s_4(P) + \delta'/2 .$$

Either way, $\mod (B'_\varphi)$ exceeds $s_4(P)$ by a definite amount bounded in terms of δ' and we invoke Proposition 4 for $\phi := \varphi$ to finish the proof.

$\square$

4.3.2   Proof of Theorem 1.2

Consider the box mapping ϕ_0 introduced in the statement of Theorem 1.2. First, observe that each univalent branch b of ϕ_0 has a separation symbol $(\alpha_0, 0, \alpha_0, \alpha_0)$ with $A_2(b)$ and $A_3(b)$ degenerate. This leads to a normalized symbol with norm $\alpha_0/2$ and corrections $\lambda_1(b) = 0$ and $\lambda_2(b) = \alpha_0/2$. By Proposition 5, every box mapping $\tilde{\phi}$ obtained from ϕ_0 by arbitrary number of type I inducing steps has the separation index larger or equal to $\alpha_0/2$ and $\mod (\tilde{B}' \setminus \tilde{B}) \geq \alpha_0/8$.

To prove Theorem 1.2, we will consider a type I mapping $\phi := \phi_n$ with a separation index $\beta \geq \alpha_0/2$ and with $\mod (B' \setminus B) \geq \alpha_0/8$. Then we will show that after a fixed number of type I inducing steps a type I box mapping is derived with a separation index $\beta + \delta$ where $\delta > 0$ depends only on α_0. First, this happens if ϕ makes a close return. This follows directly from Lemmas 4.3.2 and 4.3.1. So without loss of generality we can assume that ϕ makes 10 consecutive non-close returns. Let us denote the derived mappings with ϕ_i, $i = 1, \cdots, 10$ and use ψ_i, B_i and B'_i for the central branch of ϕ_i, its domain and range, respectively. For $i > 0$ all parent domains are separated from the boundary of B'_i by annuli with modulus $\alpha_0/16$. This happens because the parent domains are preimages by ψ_{i-1}

of domains of various branches of ϕ_{i-1}, and all these, including the central domain B_{i-1}, are surrounded in B'_{i-1} by annuli with modulus $\alpha_0/8$. From Lemmas 4.1.6-4.1.8 it now follows that for $i > 1$ all univalent branches b of ϕ_i, except for the two immediate ones, satisfy

$$\mathrm{mod}\,(A'(b) \oplus A_4(b)) \geq s_4(b) + \delta_1$$

where $\delta_1 > 0$ depends only on α_0 and s_4 is a component of some normalized separation symbol with norm β. If for any i between 2 and 9 the critical value of ϕ_i is in the domain of a non-immediate branch, then Proposition 4 implies the needed increase in the separation index. So, we have reduced the situation to a sequence of "Fibonacci" returns for $i = 2, \cdots, 9$ where the critical value is always in the domain of an immediate branch.

Fibonacci returns. This case can be handled as follows. Let $\phi := \phi_i$ and $\tilde{\phi} := \phi_{i+1}$, $2 \leq i \leq 8$. Denote by $s(P)$ a normalized separation symbol for a postcritical branch P of ϕ,

Suppose now that $s_1(P) - s_2(P) \geq \nu$. We will show that if b is an immediate branch of $\tilde{\phi}$ then we have that for every $\nu, \alpha_0 > 0$ there is an $\eta > 0$ so that b has a normalized separation symbol $s(b)$ with norm β and

$$\mathrm{mod}\,(\tilde{B}' \setminus D_b) \geq s_4(b) + \eta \,. \tag{4.14}$$

Indeed, we apply Lemma 4.1.1 in the following context. Make f equal to the central branch of ϕ, U_1 equal to $A'(P) \oplus A_3(P)$ and $D'_1 := B$. Then $U_2 = (\tilde{B}' \setminus \tilde{B}) \oplus A_2(P)$ while $W_2 = A_1(P)$. So we can put $\sigma_1 = s_1(P)$ and $\sigma_2 = \mathrm{mod}\,A_2(P) \geq s_2(P)$. Since $\mathrm{mod}\,U_2 \geq \mathrm{mod}\,(\tilde{B}' \setminus \tilde{B}) \geq \alpha_0/8$ and $\sigma_1 - \sigma_2 \geq \mathrm{mod}\,A_1(P) \geq \nu$ we get a positive η which depends only on α_0 and ν so that

$$\mathrm{mod}\,(\tilde{B}' \setminus D_b) \geq \frac{s_1(P) + s_2(P)}{2} + s_1(b) + \eta \,. \tag{4.15}$$

Looking into the proof of Lemma 4.1.5 we find that

$$s_4(b) = \beta + \frac{\lambda_1(P)}{2} \qquad \text{while}$$

$$s_1(b) = \frac{\beta + \lambda_2(P)}{2} \,.$$

Taking into account that $s_1(P) = \alpha + \lambda_1(i)$ and $s_2(P) = \alpha - \lambda_2(i)$ are elements of a normalized separation symbol, we can rewrite estimate (4.15) to get (4.14).

Now we conclude the proof of Theorem 1.2. Denote by P the branch of ϕ_2 whose domain contains the critical value and apply Lemma 4.3.2 with $\phi := \phi_2$. We get two possibilities. If the boundary of B_2 is (M, ϵ)-rough with M chosen appropriately, then Lemma 4.3.1 implies that ϕ_6 shows an increase in the separation index. So let us concentrate on the other possibility, namely that the domain of P is surrounded in B_2' by an annulus with modulus at least $s_3(P) + \eta_1$ where $s_3(P)$ is a component of a normalized separation symbol with norm β and η_1 is a positive constant which depends only on α_0. If $s_4(P) - s_3(P) \leq \eta_1/2$, then Proposition 4 gives us an increased separation index for ϕ_3. So let us assume that $s_4(P) - s_3(P) = \lambda_1(P) + \lambda_2(P) > \eta_1/2$. If b is an immediate branch of ϕ_3, Lemma 4.1.5 tells us that b has a normalized separation symbol $s(b)$ with norm β and that

$$s_1(b) - s_2(b) = \lambda_1(b) + \lambda_2(b) = \frac{\lambda_1(P) + \lambda_2(P)}{2} > \frac{\eta_1}{4} \ .$$

Recall that the critical value of ϕ_3 is in the domain of an immediate branch and apply the estimate (4.14) with $\phi := \phi_3$. We get an increase of the separation index for ϕ_4. The proof of Theorem 1.2 is finished.

Chapter 5

Quasiconformal Techniques

5.1 Initial Inducing

The main objective of this section is to present a proof of Theorem 1.3. So we assume that unimodal polynomials f and $\hat{f}$ are real, topologically conjugate, the critical orbits omit the fixed points, and have *odd* periodic orbits on the real line. These are topological assumptions and if f satisfies them than its both fixed points are repelling, f has orbits with infinitely many different periods (Sharkovski's theorem) and the first return time to the restrictive interval is greater than 2.

5.1.1 Yoccoz partition

In the proof Theorem 1.3 a major issue is the choice of domains of analytic continuations of branches of initial induced maps Φ and $\hat{\Phi}$ considered in Subsection 1.3.3. Another technical problem is the construction of the branchwise equivalence. A short answer to the first issue is that domains will be chosen as pieces of the Yoccoz partition shown on Figure 5.1.

The uniformizing map. Take a polynomial f_a defined by $f_a(z) = a(1 - z^2) - 1$ with $|a - 2| < 2$.

Fact 5.1.1 *For every f_a with $|a - 2| < 2$, there is an analytic map H_a defined on $D(0, \frac{1}{100})$ which satisfies*

$$H_a(z^2) = f_a(H_a(z))$$

on its domain. In addition, $z \to \frac{1}{H_a(z)}$ fixes 0 and has derivative $-a$ at 0. Also, for every fixed $z \in D(0, \frac{1}{100})$, $H_a(z)$ is an analytic function of a on $D(2, 2)$.

The mapping H_a introduced in Fact 5.1.1 is widely known as the *Böttker coordinate*. The domain of definition of the Böttker coordinate can be extended by pull-back. Namely, on the disk $D(0, \frac{1}{10})$, the coordinate can be defined by taking the lift of H_a to the branched covers f_a in the neighborhood of infinity and $z \to z^2$ of $D(0, \frac{1}{10})$ onto $D(0, \frac{1}{100})$. This is possible to do provided that the image of 0 by f_a is not in $H_a(D(0, \frac{1}{100}))$. Then we can proceed extending the Böttker coordinate onto larger and larger subsets of $D(0, 1)$. If the critical value of f_a is in its Julia set, this process can be continued to the limit and gives a conjugacy defined on the entire $D(0, 1)$ onto the complement of the Julia set of f_a. Observe that these extensions of the Böttker coordinate still depend analytically on a when z is fixed.

Double rays.   Consider the images of rays with arguments $2\pi/3$ and $4\pi/3$ in $D(0, e^{-100})$ under H_a. This will give a Jordan arc passing through infinity. Call this arc $\mathcal{R}_a^0$. Given $\mathcal{R}_a^n$, define $\mathcal{R}_a^{n+1}$ as the connected component of $f_a^{-1}(\mathcal{R}_a^n)$ which contains $\mathcal{R}_a^n$. The union of all $\mathcal{R}_a^n$ and the fixed point q different from -1 will be called the *double ray $\mathcal{R}_a$*.

Presently, we will show that $\mathcal{R}_a$ is a Jordan curve provided a is positive and does not exceed 2 and f_a has infinitely many periodic points on the real line. As a matter of terminology, note that images by H_a of radii are called rays while images of circles centered at 0 are called equipotential curves. Also, notice that $H_a^{-1} \circ f_a \circ H_a(z) = z^2$. The fixed point q is repelling and thus accessible from the outside of the Julia set. In fact, at least two external rays land at q since the Julia set contains the interval $[-1, 1]$. f_a is a local diffeomorphism at q and thus permutes external rays landing there.

The rays corresponding to radii with arguments $2\pi/3$ and $4\pi/3$ (rational) are known to both converge to points in the Julia set (see [6]), say p_1 and p_2. The rays are conjugate and interchanged by the dynamics of f_a, hence $f_a(p_1) = p_2$ and $f_a(p_2) = p_1$. Furthermore, p_1 and p_2 are conjugate. Thus p_1 is a periodic point of period at most 2 and the eigenvalue of f_a at p_1 must be real. Hence $p_1 = p_2$ and p_1 is fixed. Moreover, since the rays are interchanged, the derivative at p_1 must be negative, and this leaves $p_1 = q$ as the only possibility. So in this case the double ray is just the union of these two rays with ∞ and q.

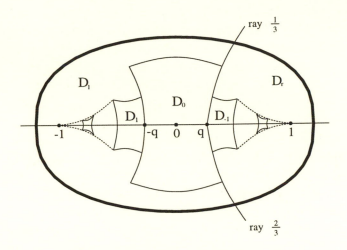

Equipotential of level 100

Figure 5.1: Yoccoz Partition. f is a unimodal quadratic polynomial with odd periods: $D_0 = f^{-1}(D_r)$, $D_i = f_l^{-i}(D_0)$, f_l is the left "lap" of f. $D_i = -D_i$, D_i tend to -1. Then take the first return map into D_0. This is a type II holomorphic box mapping. Apply filling-in to get ϕ_0.

The initial partition. Assume still that f_a is real and unimodal so that the critical value is in the Julia set. Take the image under H_a of the circle of radius e^{-100}. The bounded component of the complement of this equipotential is cut by the double rays in two pieces: D_r which does not contain 0, and D_l. Then we define $D = f_a^{-1}(D_r)$ which is contained in D_l and surrounded by another equipotential, this time the image of the circle with radius e^{-50}. Observe that $(D_r \cup D_l) \setminus \overline{D}$ is a ring domain with modulus at least 50. The boundary of $D_l \cup D_r$ is symmetric with respect to the real line.

5.1.2   Holomorphic motions and q.c. correspondence

Take a real unimodal polynomial f_a which has periodic orbits for infinitely many periods and consider a particular branch of preimages of 0. Namely, $c_n = f^{-n}(0)$ where the right lap of f_a is always used to take the preimages. Points c_n tend to q_a exponentially fast, that is $|q_a - c_n| \leq Q_1^n$ where Q_1 is independent of f_a.

Lemma 5.1.1 *Consider a real unimodal polynomial f_a which has periodic orbits with infinitely many different periods. There is a neighborhood U of a in the parameter plane so that for every $b \in U$ the critical value $f_b(0)$ does not belong to the double ray $\mathcal{R}_b$ and for every $n > 0$ there is a holomorphic function $c_n(b)$ defined on U so that $f_b^n(c_n(b)) = 0$ and $c_n(a) = c_n$ as defined above.*

Proof:

For every finite n, we can certainly find a neighborhood U_n of a so that for every $b \in U_n$ the critical value of f_b avoids $\mathcal{R}_b^n$. If this is violated, then $f_b^{n+1}(0)$ is outside of the equipotential corresponding to the circle of radius e^{-100} under H_b. For $b = a$, $f_a^{n+1}(0) \in [-1, 1]$ while the equipotential certainly contains the ball centered at 0 of radius 100 (this is because the ring domain between the Julia set and the equipotential has modulus 100, while the Julia set contains the interval $[-1, 1]$). Since both objects move holomorphically, there is an open neighborhood where they do not collide. Choose n so large that $\mathcal{R}_a^{n-1} \setminus \mathcal{R}_a^{n-2}$ is in a neighborhood of q_a mapped strictly into itself by the analytic continuation of the right lap of f_a. Then if b is allowed to change inside U_n, $\mathcal{R}_b^n$ moves holomorphically with b, thus on a smaller neighborhood U' of a we get that $\mathcal{R}_b^n \setminus \mathcal{R}_b^{n-1}$ is in a neighborhood of q_b mapped strictly into itself by the inverse branch of f_b which preserves q_b. In addition, this neighborhood may be chosen small enough not to contain $f_b(0)$. But now we see that for every $b \in U'$, $f_b(0)$ avoids the entire double ray since $\mathcal{R} \setminus \mathcal{R}_b^n$ remains in this neighborhood of q_b.

The neighborhood V' on which $c_n(b)$ are well defined is found in a similar way. As a branch of preimages of 0, $c_n(b)$ is holomorphic at b provided that $f_b^i(c_n(b)) \neq 0$ for $0 \leq i < n$. Again, it is clear that for every finite n a neighborhood V_n of a can be found so that this is satisfied for every $b \in V_n$. Pick n so large that $c_n(a)$ belongs to a neighborhood of q_a which is disjoint with 0 and mapped strictly inside itself by the inverse branch of f_a which preserves q_a. Then for b in some smaller neighborhood V', $c_n(b)$ belongs to a neighborhood of q_b which is mapped strictly inside itself by the inverse branch of f_b which preserves q_b and is disjoint with 0. Now $c_n(b)$ can all be found in the same neighborhood and hence avoid 0. So we can take $U = U' \cap V'$.

□

Lemma 5.1.2 *Consider two real unimodal polynomials f and $\hat{f}$ both with infinite sets of periods, but not necessarily topologically conjugate. Take their initial Yoccoz partitions D_l, D_r, $\hat{D}_l$ and $\hat{D}_r$, respectively. There is a constant Q independent of f and $\hat{f}$ so we can find a Q-quasiconformal homeomorphism Υ_i, $\Upsilon_i(\bar{z}) = \overline{\Upsilon_i(z)}$ on the boundary of $D_l \cup D_r$, which maps D_r onto $\hat{D}_r$ and D_l and $\hat{D}_l$ while the double ray of f goes to the double ray of $\hat{f}$. Furthermore, the functional equation*

$$(\Upsilon_i \circ f)(z) = (\hat{f} \circ \Upsilon_i)(z)$$

is satisfied for every z on the union of the borders of $D_r \cup D_l$ as well as for $z = f(0)$ and $z = c_n$.

Proof:
The tool of the proof is the λ-lemma. A weaker version of the λ-lemma originally appeared in [27] where its applications to the study of holomorphic dynamics were also exploited.

Fact 5.1.2 *Consider a family of injective mappings h_λ all defined on a set $S \in \hat{\mathbb{C}}$ and into $\hat{\mathbb{C}}$ where the parameter λ varies throughout the unit disk. Assume that h_0 is the identity and that for every fixed $z_0 \in S$, $h_\lambda(z_0)$ is a meromorphic function of λ. Then, for every $a < 1$ and $|\lambda| \leq a$ the mapping h_λ has a $\frac{1+|a|}{1-|a|}$-quasiconformal extension as an automorphism of $\hat{\mathbb{C}}$.*

Proof:
This directly follows from Theorem 1 of [2] and Theorem 1.3 of [37].

□

Take a real unimodal polynomial f_a with an infinite set of periods and its initial Yoccoz partition (D_l^a, D_r^a). The set S is chosen to contain the complement of $\overline{D_l^a \cup D_r^a}$, the entire double ray, $f_a(0)$ and $c_n(a)$. Choose a round disk U^a centered at a and contained in the neighborhood U obtained from Lemma 5.1.1 for this particular a. After an affine change of variables, U^a becomes the parametric unit disk in Fact 5.1.2. We will use the name b for the parameter in U^a. The branch $c_n(b)$ is well-defined and holomorphic on U^a while $f_b(0)$ avoids the double ray. The mappings h_b are defined as follows. In the complement of $\overline{D_l^a \cup D_r^a}$, set $h_b = H_b \circ H_a^{-1}$. On the remaining part of the double ray, h_b is defined to satisfy

$$h_b = f_b^{-1} \circ h_b \circ f_a$$

where the inverse branch which fixes q_b is chosen. This branch is well-defined on the double ray since the critical value is far away. Finally, h_b sends $f_a(0)$ to $f_b(0)$ and $c_n(a)$ to $c_n(b)$. If b is real, $h_b(\overline{z}) = \overline{h_b(z)}$ on the boundary of $D_l^a \cup D_r^b$.

Let us check the assumptions of Fact 5.1.2 (with the implicit rescaling of b). We noted earlier that H_b moves analytically outside of the equipotential corresponding to the circle with radius $1/100$, and this clearly contains the complement of $\overline{D_l^a \cup D_r^a}$. It is then obvious that h_b also moves analytically on the remaining part of S. Also h_a is the identity. Each h_b is injective on the union of the double ray and $\overline{D_l^a \cup D_r^a}$. $f_b(0)$ does not hit that region by the choice of U^a. Neither does $c_n(b)$ as $f_b(0)$ is in its forward orbit. Finally, $f_b(0)$ and $c_n(b)$ do not collide as this would mean $f_b^{n-1}(c_n(b)) = 0$ and thus $f_b(0) = 0$. Let $U_{1/2}^a$ be the disk centered at a with the radius half as large as U^a. By Fact 5.1.2, the claim of Lemma 5.1.2 is fulfilled with $f = f_a$ and $\hat{f} = f_b$ where $b \in U_{1/2}^a$. To finish the proof, choose a finite subcover of the cover of the closed interval in the parametric space corresponding to real unimodal polynomials with infinite sets of periods by neighborhoods $U_{1/2}^a$.

$\square$

The following is a corollary from Lemma 5.1.2.

Lemma 5.1.3 *Let f_a be a real unimodal polynomial with an infinite set of periods for its periodic orbits. Then the double ray $\mathcal{R}_a$ is Q-quasiconformal.*

Proof:

Apply Lemma 5.1.2 with $f = f_a$ and $\hat{f} = f_2$. It is enough to see that the double ray $\mathcal{R}_2$ is quasiconformal. For the Chebyshev polynomial f_2 the Julia set is $[-1, 1]$ and the uniformizing map is known explicitly to be $H_2(z) = -\frac{1}{2}(z + z^{-1})$. A simple calculation shows that the double ray $\mathcal{R}_a$ is the image of the line $\{x + i\frac{\pi}{3} : x \in \mathbb{R}\}$ by $-\cosh z$. On this line, the derivative $-\sinh z$ satisfies $\arg\sinh z \in (-\frac{\pi}{3}, -\frac{2\pi}{3})$, and hence the length of the double ray between two points is bounded by twice the difference between the imaginary parts. By Fact 2.3.2, the part of the double ray with infinity removed is a quasiconformal arc, but in a neighborhood of infinity the double ray is also quasiconformal since it is the image of the radii with arguments $2\pi/3$ and $4\pi/3$ by the smooth map H_2. Since the double ray can be covered by two quasiconformal arcs, it is quasiconformal itself by Theorem 8.7, page 103 of [25].

$\square$

Next, we will look a little closer at the geometry of the piece D^a of the Yoccoz partition. Recall that $D^a = f_a^{-1}(D_r)$.

Lemma 5.1.4 *Let f^a be a real unimodal polynomial with an infinite set of periods. The piece D^a is symmetric with respect to the real line and bounded by a finite union of quasiconformal Jordan arcs. Moreover, there is a constant $\alpha > 0$ independent of f_a so that $D_a \supset D(\alpha, (-q_a, q_a))$ where $D(\cdot, \cdot)$ is a geodesic neighborhood, see Definition 2.1.2.*

Proof:

The symmetry is clear since D_r^a is symmetric and f_a is a real polynomial. The border of D^a consists of two pieces of an equipotential, certainly quasiconformal as an image of a circle by a smooth map, see Fact 4.2.3, a piece of the double ray $\mathcal{R}_a$ and a piece of the mirror image of $\mathcal{R}_a$ across the imaginary axis. These are quasiconformal arcs by Lemma 5.1.3. Suppose that a point z belongs to $\mathcal{R}_a \cap D(\alpha, (-q_a, q_a))$. Then so does its conjugate point $\bar{z}$. The diameter of the stretch of $\mathcal{R}_a$ between z and $\bar{z}$ is at least $z - q_a$. Clearly, $\frac{|z - q_a|}{|z - \bar{z}|}$ goes to infinity as α decreases to 0. On the other hand, by Lemma 5.1.3 and Fact 2.3.2, this ratio is bounded by a constant. The same argument applies to z in the mirror ray passing through

$-q_a$. The part of the boundary of D^a contained in the equipotential is far away, certainly outside of the circle of radius 100 around 0. Hence for α less than some positive constant the boundary of D^a does not enter $D(\alpha, (-q_a, q_a))$.

$\square$

Lemma 5.1.5 *Take two real unimodal polynomials f and $\hat{f}$ which are topologically conjugate and have periodic orbits of odd periods. Let Υ_i denote a mapping constructed by Lemma 5.1.2 for some n. For every point $z_0 \in (-q, q)$ there is a Q-quasiconformal homeomorphism Υ_0 and a choice of Υ_i with the following properties:*

- *$\Upsilon_0(\bar{z}) = \overline{\Upsilon_i(z)}$ on the boundary of D and $\Upsilon_0 = \Upsilon_i$ except on D_l,*

- *the functional equation*

$$(\hat{f} \circ \Upsilon_0)(z) = (\Upsilon_i \circ f)(z)$$

  is satisfied for z on the boundary of D,

- *Υ_0 at z_0 coincides with the topological conjugacy between f and $\hat{f}$.*

The bound Q is a constant independent of f, $\hat{f}$ or z_0.

Proof:
First choose a number n which has to be entered in Lemma 5.1.2 in order to get Υ_i. Choose n the largest so that $c_{n+1} \leq |z_0|$. Apply Lemma 5.1.2 to f, $\hat{f}$ and n in order to get Υ_i. Then consider Υ_1 defined as the lift of Υ_i to the branched covers f and $\hat{f}$. The lifting is topologically admissible because $\Upsilon_i(f(0)) = \hat{f}(0)$. Among two possible liftings, make Υ_1 the one which maps $\mathcal{R}$ on $\hat{\mathcal{R}}$. Υ_1 coincides with Υ_i on the border of D_l because the functional equation $\hat{f} \circ \Upsilon_i = \Upsilon_i \circ f$ was satisfied there (see Lemma 5.1.2.) Now set Υ_2 equal to Υ_1 on D_l and Υ_i elsewhere. The mapping Υ_2 is a homeomorphism of the whole complex plane and its quasiconformal maximal dilatation bound is as for Υ_i in the complement of the boundary of D_l. By Lemma 5.1.4 this boundary is a finite union of quasiconformal Jordan arcs. Each of these arcs is removable (a direct consequence of Theorem 8.3 page

45 of [25]), and the maximal dilatation is bounded everywhere as for Υ_i. Υ_2 satisfies most properties claimed for Υ_0 with the sole exception of being equal to the topological conjugacy at z_0. Instead, we have $\Upsilon_2(c_{n+1}) = \hat{c}_{n+1}$ and $\Upsilon_2(-c_{n+1}) = -\hat{c}_{n+1}$. Suppose that $z_0 > 0$. Then $z_0 \in [c_{n+1}, c_{n+3})$. Since points c_k are mapped to $\hat{c}_k$ by the conjugacy, the conjugate point $\hat{z}_0$ belongs to $[\hat{c}_{n+1}, \hat{c}_{n+3})$. Points c_k converge to q at a universal exponential rate, and thus the ratio

$$\frac{q - c_{n+1}}{q - z_0} \leq Q_1$$

where Q_1 is a constant independent of f, $\hat{f}$ and z_0. This means that for any $\pi > \alpha > 0$, the distance from c_{n+1} to z_0 is bounded by a constant in the standard Poincaré metric of $D(\alpha, (-q, q))$. In view of Lemma 5.1.4, the distance between these points is also bounded by a constant in the Poincaré metric of D_l. The same argument shows that the distance from $\hat{z}_0$ to $\hat{c}_{n+1}$ is bounded by a constant in the Poincaré metric of $\hat{D}_l$. Since Υ_2 is quasiconformal, we conclude that the Poincaré distance in $\hat{D}_l$ from $\hat{c}_{n+1}$ to $\Upsilon_2(z_0)$ is bounded by a constant. So the Poincaré distance in $\hat{D}_l$ from $\Upsilon_2(z_0)$ to $\hat{z}_0$ is bounded by a constant. By Lemma 5.2.3 we can find a Q_2-quasiconformal homeomorphism Υ_3 which is the identity outside of D and maps $\Upsilon_2(z_0)$ to $\hat{z}_0$. Here Q_2 is still a constant independent of f, $\hat{f}$ or z_0. Then we can set $\Upsilon_0 = \Upsilon_3 \circ \Upsilon_2$ and the Lemma is proved. The case of $z_0 < 0$ is symmetrical when one uses the fact that $\Upsilon_2(-c_{n+1}) = -\hat{c}_{n+1}$. Clearly, $\Upsilon_0(\bar{z}) = \overline{\Upsilon_0(z)}$ on the boundary of D since Υ_0 coincide there with the pull-back of Υ_i.

□

Inner staircases. Let us take a real unimodal polynomial f_a with an infinite set of periods. Recall the initial Yoccoz construction of the domain D^a which is a topological disk based on $(-q_a, q_a)$. Denote $-q_a = q_0$ and take the sequence of preimages of q_0 by the left lap of f_a. Thus, $q_n = f_a(q_{n+1})$. The holomorphic branch of f_a^{-1} which is the analytic continuation of the left lap gives a sequence of topological disks D_n so that $D_0 = D^a$ and $f_a^n(D_n) = D_0$. Also, D_n is based on (q_n, q_{n-1}). The sequence (q_n) is called the *left inner staircase* of f_a and the sequence $(-q_n)$ is the *right inner staircase*.

Now take two topologically real unimodal polynomials f and $\hat{f}$, not necessarily topologically conjugate, which have periodic orbits with odd periods. We will construct a quasiconformal correspondence between their inner staircases, also respecting the topological disks D_n. If f and $\hat{f}$ are conjugate, we apply Lemma 5.1.5 where z_0 is chosen as the first return of the forward critical orbit of f into $(-q, q)$. This gives us a Q_1-quasiconformal homeomorphism Υ_0. In the general case, we can always take Υ_0 equal to the Υ_i obtained by Lemma 5.1.2.

Then launch an inductive process as follows. Given Υ_n replace Υ_n on D_l by the lift of Υ_n to the branched covers f and $\hat{f}$. The lift is chosen so that q gets mapped to $\hat{q}$. The mapping obtained in this way is Υ_{n+1}. This construction is feasible provided that $\Upsilon_n(f(0)) = \hat{f}(0)$ and gives a homeomorphism provided that the functional equation $\hat{f} \circ \Upsilon_n = \Upsilon_n \circ f$ is satisfied on the boundary of D_l. Both conditions are satisfied since Υ_n remains fixed outside of D_l and both conditions are satisfied by Υ_0. Thus Υ_n form a sequence of Q_1-quasiconformal homeomorphisms of the plane. Observe that Υ_n maps every D_i, $i \leq n$, onto $\hat{D}_i$. Since mappings Υ_n are fixed except of D_l, they form an equicontinuous family on the sphere. Hence there is C^0 limit, called Υ_∞ which is also Q_1-quasiconformal. We have $\Upsilon_\infty(D_n) = \hat{D}_n$ for every n. Moreover, the functional equation

$$\hat{f} \circ \Upsilon_\infty = \Upsilon_\infty \circ f$$

is satisfied on the boundary of each D_n, D_r and D_l, and still $\Upsilon_\infty = \Upsilon_0$ outside of D_l. Therefore, $\Upsilon_\infty(f(0)) = \hat{f}(0)$. If f and $\hat{f}$ were topologically conjugate, as the consequence of our initial choice of z_0, we also have $\Upsilon_\infty(f^2(0)) = \hat{f}^2(0)$. Now let us take the lift Υ' of Υ_∞ by f and $\hat{f}$ chosen so that $\Upsilon'(q) = \hat{q}$. This time, however, we replace Υ_∞ with Υ' not only on D_l, but on D_r as well. Still, we get a Q_1-quasiconformal mapping Υ_1'. Υ_1' also establishes a correspondence between the right inner staircases and the disks which are symmetric to D_n and $\hat{D}_n$, respectively. Also, we have

$$\hat{f} \circ \Upsilon_1' = \Upsilon_1' \circ f$$

on the boundary of any disk D_n or $D_{-n} := -D_n$ as well as at $f(0)$. By the construction, the disks D_n, D_0, and D_{-n} are symmetric with respect to the real line and $\Upsilon_1'(\bar{z}) = \overline{\Upsilon_1'(z)}$ on their boundaries.

In particular, we have proved the following fact:

Fact 5.1.3 *Let f and $\hat{f}$ be real unimodal polynomials with odd periods. Then we have a Q-quasiconformal homeomorphism Υ'_1 of the plane, where Q is a constant, which maps D_r for f onto the corresponding $\hat{D}_r$ for $\hat{f}$, the disks D_{-n} of the right inner staircase of f onto the corresponding disks of the right inner staircase of $\hat{f}$, and coincides with the topological conjugacy on the border of D_r and the borders of D_{-n} for $n = 1, \cdots$. In addition, if f and $\hat{f}$ are topologically conjugate, then $\Upsilon'_1(f(0)) = \hat{f}(0)$.*

As a corollary (see Lemma 2.3.1), we get

Fact 5.1.4 *There is a constant Q' so that for every pair of quadratic polynomials f and $\hat{f}$, both with infinite sets of periods, there is a Q'-quasi-symmetric homeomorphism μ of $[-1,1]$ onto itself which maps the inner staircases of f onto the inner staircases of $\hat{f}$ (i.e. $\mu(q_n) = \hat{q}_n$ and $\mu(-q_n) = -\hat{q}_n$ for every n.)*

Proof:
The map Υ' constructed above send the inner staircases of f onto the inner staircases of $\hat{f}$, but may fail to preserve the real interval. But it can be modified on each disk D_n to preserve the real line by Lemma 2.3.1.

$\square$

Proof of Theorem 1.3.   Let us build the induced mapping ϕ first. Returning to the inner staircases, we can construct a holomorphic induced map ϕ_1 defined to be f on D_0 and f^n on D_n and D_{-n}. Then $\phi_0 = \phi_1 \circ f$ provides ϕ_0 that is immediately seen to be a type II box mapping. The branchwise equivalence Υ^0 between ϕ_0 and $\hat{\phi}_0$ is obtained by lifting Υ' to the branched covers f and $\hat{f}$ so that $\Upsilon^0(q) = \hat{q}$. Observe that Υ^0 is quasiconformal with the bound for dilation independent of f and $\hat{f}$. Also, Υ' commutes with the complex conjugation except on the domain of ϕ_1. However, on these domains we can apply Lemma 2.3.1 to make them symmetric at the cost of increasing the maximal dilatation at most twice. Then the holomorphic extensions of the initial type I box mappings, that is ϕ

and $\hat{\phi}$, are obtained by filling-in with the branchwise equivalence derived from the corresponding pull-back (see Section 5.2). This gives a branchwise equivalence between ϕ and $\hat{\phi}$ which is quasiconformal with the same bound independent of f and $\hat{f}$.

It remains to verify the moduli condition for ϕ. Suppose that $f(0) \in D_{-n}$. Since f has odd periods, then $n > 1$. Then

$$\mathrm{mod}\,(B' \setminus B) = \frac{1}{2}\mathrm{mod}\,(D_r \setminus D_{-n})\,.$$

Now apply Fact 5.1.3 to f with $\hat{f}$ equal to some fixed polynomial, for example the Chebyshev polynomial. Since Υ'_1 is Q-quasiconformal, it follows that $\mathrm{mod}\,(D_r \setminus D_{-n})$ as it least Q^{-1} times the analogous modulus for $\hat{f}$, that is a fixed positive number.

5.2   Quasiconformal Pull-back

5.2.1   Definition of pull-back

We assume here that the reader is familiar with definitions from Section 1.3 regarding branchwise equivalences, their similarity classes and critical consistence.

Pull-back corresponding to filling-in.   Consider two holomorphic type II box mappings, ϕ and $\hat{\phi}$ with a quasiconformal branchwise equivalence v between them. Assume that the sets S and $\hat{S}$ are totally disconnected, can be iterated by univalent branches of respective mappings forever, and $B \cap S = \emptyset$ and $\hat{B} \cap \hat{S} = \emptyset$.

Let ϕ_∞ and $\hat{\phi}_\infty$ be the corresponding type I box mappings obtained by filling-in. A branchwise equivalence v_∞ between ϕ_∞ and $\hat{\phi}_\infty$ can be obtained as the limit of a pull-back process.

Let U_0 denote the union of domains of all branches of ϕ that map univalently onto B', and $\hat{U}_0$ be analogously defined for $\hat{\phi}$. Then v_1 on U_0 is uniquely determined by the functional equation

$$v \circ \phi = \hat{\phi} \circ v_1\,.$$

Notice that $v_1 = v$ on the border of U_0. Hence, we can define v_1 on the whole plane by setting it equal to v outside of U.

Now consider the set U_1 where the map

$$\phi_1 := \phi_{|U_0} \circ \phi_{|U_0}$$

is defined. By analogy, we have $\hat{U}_1$ and $\hat{\phi}_1$. On the border of U_1, v_1 satisfies the functional equation

$$v_1 \circ \phi_1 = \hat{\phi}_1 \circ v_1 .$$

Hence v_2 can be defined by the condition

$$v_1 \circ \phi_1 = \hat{\phi}_1 \circ v_2$$

on U_1 and equal to v_1 elsewhere. This process can be inductively continued to infinity. Observe that the sequence $v_k, v_{k+1}, \cdots$ stabilizes outside of U_k. On the other hand, $\bigcap_{k=0}^{\infty} U_k = S$ and $\bigcap_{k=0}^{\infty} \hat{U}_k = \hat{S}$. So, by the assumption that S and $\hat{S}$ are totally disconnected, the sequence v_k has a well-defined limit v_∞. By the construction, v_∞ is a branchwise equivalence between type I box maps ϕ_∞ and $\hat{\phi}_\infty$. We will call v_∞ the *pull-back of v corresponding to filling-in*. Notice that the whole process preserves the similarity relation, and so pull-back corresponding to filling-in is well-defined on similarity classes of branchwise equivalences.

We will show that v_∞ is quasiconformal and has the same dilatation as v. Moreover, somewhat paradoxically, the pull-back can help us to get rid of large dilatation.

Vanishing dilatation. Let H be a quasiconformal homeomorphism in the plane. The *Beltrami bound* of H on some set $S \subset \mathbb{C}$ is equal to L^∞ norm of $\frac{H_{\bar{z}}}{H_z}\chi_S$.

Proposition 6 *Let ϕ and $\hat{\phi}$ be type II holomorphic box mappings with a quasiconformal branchwise equivalence v. The unions of domains of univalent branches of ϕ and $\hat{\phi}$ are respectively denoted by U and $\hat{U}$. Assume that the sets S and $\hat{S}$ of points that stay in U or $\hat{U}$, respectively, forever under iteration are totally disconnected and of zero Lebesgue measure.*

Then for any compact set L contained in U, the Beltrami bound of the branchwise equivalence v_∞ obtained by the pull-back corresponding to filling-in is bounded by the Beltrami bound of v outside of L.

Let κ stand for the Beltrami bound of v globally. Observe that every v_k is quasiconformal with the same Beltrami bound κ as v. The proof follows by induction. To pass from v_{k-1} to v_k we replace v_{k-1} by an appropriate pull-back of v_{k-1} on the set U_{k-1}. Hence, v_k has Beltrami bound not larger than κ on U_{k-1} as well as on the complement of the closure of U_k. But it is not immediately clear that it is quasiconformal on $\mathbb{C}$. The problem is solved by a lemma based on the work of Lipman Bers, which was pointed up to us by E. de Faria.

Lemma 5.2.1 *Let H be a K_1-quasiconformal automorphism of the Riemann sphere. Suppose that G is another such automorphism, assumed only to be topological. However, on some open set W, G is K_2-quasiconformal, while on the complement of W it is equal to H. Then, G is $\max(K_1, K_2)$-quasiconformal.*

Proof:
Lemma 1 in [1] states that

Fact 5.2.1 *If $D \subset \hat{\mathbb{C}}$ is open, f is a topological automorphism of $\hat{\mathbb{C}}$ such that $f_{|D}$ is quasiconformal and $f_{|\hat{\mathbb{C}} \backslash D} = id$, then f is quasiconformal everywhere and $\mu_{|\hat{\mathbb{C}} \backslash D} = 0$.*

Apply Fact 5.2.1 with $f := G \circ H^{-1}$ and $D := H(W)$. Then $G = f \circ H$ is quasiconformal with μ equal to μ_G on W and μ_H on the complement.

$\square$

Lemma 5.2.1 applied with $W = U_{k-1}$ yields that v_k is quasiconformal with a Beltrami bound κ.

Let K be the Beltrami bound of v outside of L and L_k consist of those points in U_k which are mapped into L by ϕ^k. By induction, the maximal dilatation of v_k outside of $L_k \subset U_k$ is bounded by K.

It is not hard to see that

$$\sum_{k=1}^{\infty} \text{area}\, U_k < \infty .$$

Indeed, let D be a connected component of U_k. The union of L and $\overline{B}$ is compact and contained in B'. The distortion of ϕ^k on

$D \cap \phi^{-k}(L \cup \overline{B})$ is bounded independently of k and a choice of D. So,

$$\text{area } L_k \cap D < Q \cdot \text{area } \phi^{-k}(B) \cap D \ .$$

But sets $\phi^{-k}(\overline{B}) \cap D$ are disjoint for various k and D and contained in a compact set of $\mathbb{C}$. So the sum of the areas of L_k is finite, in particular area $L_k \to 0$.

To prove that K is the Beltrami bound for v_∞, we show that v is $\frac{K+1}{1-K}$-quasiconformal. By Theorem IV.3.3 in [25], this means that for every rectangle R we have $\text{mod } v_\infty(R) \le \frac{K+1}{1-K} \text{ mod } R$. It is also known, see formula V.6.7 in [25], that for every quasiconformal H and every rectangle R contained in the domain of H,

$$\text{mod } H(R) \le \frac{\int_R \frac{1+|\mu_H|}{1-|\mu_H|} \, dz d\bar{z}}{\text{area } R} \text{ mod } R \ .$$

Fixing R and using this formula for $H := v_k$, we get

$$\text{mod } v_k(R) \le \left[\frac{K+1}{1-K} + \frac{\kappa+1}{1-\kappa} \frac{\text{area } L_k}{\text{area } R} \right] \text{ mod } R \ .$$

Since area $L_k \to 0$,

$$\limsup_{k \to \infty} \text{mod } v_k(R) \le \frac{K+1}{1-K} \text{ mod } R \ ,$$

but

$$\lim_{k \to \infty} \text{mod } v_k(R) = \text{mod } v_\infty(R)$$

by the theorem about continuity of the modulus with respect to the Hausdorff metric, see section I.4.9 in [25].

This concludes the proof of Proposition 6.

Pull-back corresponding to critical filling. Now suppose that two type I complex box mappings ϕ and $\hat{\phi}$ are given with a critically consistent branchwise equivalence v. Recall here that the critical consistency condition means $\hat{\phi}(0) = v(\phi(0))$. Assume that the critical values are in the domains of the respective maps so that critical filling is well-defined. This will give type II box mappings ϕ' and $\hat{\phi}'$, respectively. To obtain a branchwise equivalence Υ between ϕ' and

$\hat{\phi}'$, define Υ first on the central domain of ϕ by lifting v, that is, by the condition

$$v \circ \phi = \hat{\phi} \circ \Upsilon .$$

The lifting is made possible by the critical consistency condition. Among two possible liftings, choose one that equals v on the border of the central domain. Then extend Υ to the whole plane by v. We will call this process *the pull-back corresponding to critical filling*.

Note that the pull-back corresponding to critical filling is well-defined on similarity classes. Also, if v is Q-quasiconformal, then so is Υ, from Lemma 5.2.1.

Pull-back and inducing steps. Since the simple inducing step is a composition of filling-in and critical filling, the pull-back corresponding to a simple inducing step is also defined. It is more natural to talk about pull-back of branchwise equivalences between type I box mappings, because critical consistency is required for the branchwise equivalence between such maps. For type II box mappings, we would need critical consistency between the branchwise equivalences obtained by the pull-back corresponding to filling-in.

5.2.2   Maximal dilatation and the pull-back

For quasiconformal critically consistent branchwise equivalences, the pull-back does not increase the maximal dilatation. But critical consistency of individual branchwise equivalences cannot be expected to be preserved by pull-back. Only critical consistency of similarity classes can be expected to hold, thanks to the fact that we are working with topologically conjugate maps. Within a critically consistent similarity class, a *correction* will be needed to make an individual representative critically consistent. The effect of this correction of the maximal dilatation needs to be estimated. The correction is illustrated on Figure 1.1.

The cost of correcting. Now we address the following problem. Suppose that two type I box mappings are given with a similarity class $[\Upsilon]$ of branchwise equivalence that is critically consistent for a type I inducing step. If this class contains a K-quasiconformal representative, what is the bound we can give for the maximal dilatation of a representative of the similarity class obtained by pulling $[\Upsilon]$ back? As we argued before, the increase of maximal dilatation will only be due to corrections needed to ensure critical consistency. The key to getting estimates is in assuming that Υ is critically consistent for a somewhat longer sequence of simple inducing steps. The answer is given by Proposition 1 which was stated in Section 1.3. A striking feature of the estimate in Proposition 1 is that the quasiconformal estimate only depends on the geometry of the phase space of the "hatted" mappings.

A few estimates.   Recall that the standard Poicaré metric on the unit disk is chosen to be equal to the Euclidean metric at the origin, that is its element is given by

$$d\rho = \frac{|dz|}{1 - |z|^2} .$$

Lemma 5.2.2 *Let D and S be a topological disks, with $D \supset \overline{S}$. Denote by v the modulus of the annulus $B \setminus S$.*

Then, for every $v > 0$, there is Q so that the diameter of S in the Poincaré metric of D is bounded by Q. Furthermore, if $v \geq \log 8$, then one can set $Q = Q_1 \exp(-v)$ with $Q_1 = \frac{4}{3}$.

Proof:
We can assume without loss of generality that D is the unit disk. Then we pick two points x and y inside S. By applying a Poincaré isometry we can assume that $x = 0$ and $y = \epsilon$. We will bound ϵ in terms of v. By Grötzsch's module theorem (see [25], page 54), $v \leq \mu(\epsilon)$ where $\mu(\epsilon)$ is the modulus of the annulus $D(0,1) \setminus [0, \epsilon]$. Clearly $\mu(\epsilon)$ is a decreasing function that tends to 0 as ϵ goes to 1, thus for every $v > 0$, $\epsilon \leq \mu^{-1}(v) < 1$. The Poincaré distance between x and y is bounded in terms of ϵ, so the first assertion follows.

Also, by estimate on page 61, ibidem, $\mu(\epsilon) < \log \frac{4}{\epsilon}$. Hence

$$\epsilon < 4 \exp(-v) .$$

If $v \geq \log 8$ this gives $\epsilon < 1/2$ and the Poincaré distance between x and y is bounded by $Q_1 \epsilon$ where Q_1 is a constant.

$\square$

Lemma 5.2.3 *Let D be a Jordan domain. Suppose that $x, y \in D$ and the distance between x and y in the Poincaré metric of D is d. Then, there is a homeomorphism H of the plane which is the identity outside of D, satisfies $H(x) = y$, and is e^d-quasiconformal.*

Proof:

Consider the univalent mapping M which transforms D onto the upper half plane so that $M(x) = i$ and $M(y) = ai$ for $a = e^d$. Define the mapping L from the upper half-plane into itself by

$$L(x + iy) := x + iay .$$

The mapping L is e^d-quasiconformal. Then, define

$$H = M^{-1} \circ L \circ M .$$

Observe that H extends continuously to the complement of D as the identity. We quote Lemma 5.2.1 to conclude the proof.

□

Note that Lemma 5.2.3 can be used if D is just a topological disk, since we can decrease D to a region delimited by a smooth Jordan curve that goes inside D very close to the boundary. The bound will suffer only arbitrarily little.

Proof of Proposition 1. We present the proof in a sequence of steps.

Step I. Suppose that the escaping time of ϕ is E. Look at $c := \phi^E(0)$. In other words, c is the first point of the orbit of 0 by the central branch of ϕ which is outside of the central domain. Similarly, we define $\hat{c} := \hat{\phi}^E(0)$. Let D be the connected component of the domain of ϕ which contains c. Define $\hat{D}$ analogously. By the critical consistency assumption, $\Upsilon(c) \in \hat{D}$.

We construct a mapping Υ_a as follows. Choose the smooth curve w which splits the annulus $\hat{B}' \setminus \hat{D}$ into the union of two nesting annuli, each with modulus exactly half as large. Pick a quasiconformal mapping G_1 which is the identity on the unbounded component of the complement of w while it sends $\Upsilon(c)$ to $\hat{c}$. By Lemmas 5.2.2 and 5.2.3, such G_1 exists with a dilatation bound L_1

depending only on ϵ, and if $\epsilon \geq 2\log 8$, then a specific bound by $L_1 = \exp(Q_1 \exp(-\frac{\epsilon}{2}))$ is possible with Q_1 set equal to the constant Q_1 of Lemma 5.2.2. We put $\Upsilon_a := G_1 \circ \Upsilon$.

Step II. Construct the mapping Υ_p by pulling Υ_a back. That is, Υ_p acting from B to $\hat{B}$ is defined to be the lifting of Υ_a from B' to $\hat{B}'$ by the branched covers ϕ and $\hat{\phi}$. The lift is chosen so that Υ_p and Υ match continuously at the boundary of B.

Let ϕ_p and $\hat{\phi}_p$ be the type II holomorphic box mappings obtained from ϕ and $\hat{\phi}$, respectively, by an inducing Step B. Next, let Υ_b be any representative of the similarity class of branchwise equivalences between ϕ_p and $\hat{\phi}_p$ derived by the pull-back from $[\Upsilon]$.

Next, let d be equal to the image of c by the univalent branch defined on D. Define $\hat{d}$ in the same way as the image of $\hat{c}$ by the univalent branch defined on $\hat{D}$. By the critical consistency condition, $\hat{d}$ and $\Upsilon_b(d)$ belong to the same connected component of the domain of $\hat{\phi}_p$.

Step III. We will estimate Δ defined to be the distance in the standard Poincaré metric on $\hat{B}$ from $\Upsilon_p(d)$ to $\hat{d}$. Denote by W the preimage by the central branch of $\hat{\phi}$ of the bounded connected component of the complement of w.

Assume first that $\hat{d}$ belongs to a connected component of the domain of $\hat{\phi}_p$ which is disjoint from W. Then Υ_p^{-1} and Υ_b^{-1} are the same on the border of this component, and so the critical consistency assumption implies that $\Upsilon_p(d)$ is in the same connected component. This connected component is separated from the complement of $\hat{B}$ by an annulus with modulus at least $\frac{\epsilon}{2}$. In this case, Lemma 5.2.2 gives that Δ is bounded from above in terms of ϵ, and if $\epsilon \geq 2\log 8$, then $\Delta \leq \Delta_1 := Q_1 \exp(-\frac{\epsilon}{2})$ where Q_1 is a constant.

Assume now that $\hat{d}$ belongs to a connected component of the domain of $\hat{\phi}_p$ which intersects W. Then $\Upsilon_p(d)$ is somewhere in the union of this connected component and W. The first subcase was already considered, so let us assume that $\Upsilon_p(d) \in W$. Pick an auxiliary point z in the intersection of the connected component of the domain of $\hat{\phi}_p$ which contains $\hat{d}$ and W. We will estimate Δ_2 defined to be the distance from z to $\Upsilon_p(d)$ in the standard Poincaré metric on $\hat{B}$. Then Δ_2 increased by Δ_1 found in the preceding case will give us the final estimate. The quantity Δ_2 is easily estimated from Lemma 5.2.2 because W is surrounded inside $\hat{B}$ by an annulus with modulus $\frac{\epsilon}{4}$. Hence, Δ_2 is bounded in terms of ϵ_2, and if $\epsilon \geq 4\log 8$, we can set $\Delta_2 \leq Q_1 \exp(-\frac{\epsilon}{4})$.

Thus, our final answer is that Δ is bounded in terms of ϵ, and if $\epsilon \geq 4\log 8$, then

$$\Delta \leq 2Q_1 \exp(-\frac{\epsilon}{4}) . \tag{5.1}$$

Step IV. In this step we will construct a critically consistent branchwise equivalence Υ_m which is the same as Υ outside of D, but maps c to $\hat{c}$. First, look for a quasiconformal mapping G_2 which is the identity outside of $\hat{B}$ and maps $\Upsilon_p(d)$ to $\hat{d}$. By estimate (5.1) and Lemma 5.2.3, an L_2-quasiconformal mapping G_2 with these properties exists, where L_2 is bounded in terms of ϵ, and if $\epsilon \geq 4\log 8$, then one can set

$$L_2 = \exp(2Q_1 \exp(-\frac{\epsilon}{4})) .$$

Now Υ_m is easy to construct. Let ζ denote the univalent branch of ϕ defined on D, and $\hat{\zeta}$ be the univalent branch of $\hat{\phi}$ defined on $\hat{D}$. Then set $\Upsilon_m = \hat{\zeta}^{-1} \circ G_2 \circ \Upsilon_p \circ \zeta$ inside D, and Υ elsewhere. Since Υ_p matched continuously with Υ along the border of B, this is a continuous map, and hence $L_2 L_1 K$-quasiconformal.

Step V. We are ready to finish the proof by pulling Υ_m back. In the case when ϕ makes a non-close return, it is evidently possible since Υ is critically consistent and will not entail any further growth of quasiconformal dilatation. The situation is more subtle when ϕ makes a close return because Υ_m is not critically consistent for the intermediate simple inducing steps. So, build Υ^1 quasiconformal so that $\Upsilon^1(\phi(0)) = \hat{\phi}(0)$ and $\Upsilon^1 = \Upsilon_m$ outside of B. The construction of Υ^1 is possible as long as we claim no specific estimate on quasiconformal dilatation of Υ_1 on B. Then pull Υ_1 back to get Υ_m^2. Inductively, given Υ_m^i with $i < E$, get Υ^i by changing Υ_m^i quasiconformally inside the $i-1$-st preimage of B by the central branch of ϕ in order to satisfy critical consistency, and then pull Υ^i back to get Υ_m^{i+1}.

In this way we eventually arrive at Υ_m^E. Now the critical value $\phi(0)$ is in the appropriate preimage of D, and in fact is mapped by Υ_m^E onto $\hat{\phi}(0)$ just because Υ_m sent c to $\hat{c}$ and critical consistency was assumed. Although Υ_m^E is quasiconformal, we have no specific bound on its dilatation everywhere. However, outside of the $E-1$-st preimage of B by ϕ the bound by $L_2 L_1 K$ is still valid. Next, pull Υ_m^E back again for one more simple inducing step. No problem arises for the pull-back associated with critical filling because Υ_m^E was already critically consistent. The resulting branchwise equivalence is $L_2 L_1 K$-quasiconformal except on the domains of two immediate branches of the resulting type II box map. Next, we apply filling-in and observe that the badly estimated dilatation disappears by Proposition 6. The assumptions about the limit S are easily verified. Since every domain of a univalent branch of the box mapping is separated from the boundary of B' by an annulus of definite modulus, it follows that diameters of connected components of sets U_k go to 0 at a uniform rate with k.

This ends the construction of the $L_2 L_1 K$-quasiconformal representative of the similarity class obtained by pull-back from $[\Upsilon]$. In Proposition 1, we need to set $Q := L_2 L_1$. The numbers L_1 and L_2 were estimated in Steps I and IV, respectively. Inspection of these estimates shows that both are bounded in terms of ϵ, and when $\epsilon \geq 4 \log 8$, then we can set

$$L_1 L_2 \leq Q := \exp(3Q_1 \exp(-\frac{\epsilon}{4})) \, .$$

This concludes the proof of Proposition 1.

5.3 Gluing Quasiconformal Maps

5.3.1 Quasiconformal mappings on ring domains

Interpolation across a ring domain   By analogy with analytic maps, we call a quasiconformal map $\Upsilon : \mathbb{C} \to \mathbb{C}$ *real* on the domain G if the image of $\mathbb{R} \cap G$ is real.

Here is an important technical lemma.

Lemma 5.3.1 *Let G be a ring domain bounded by two Jordan curves α and β so that β is inside the bounded component of the complement of α. Let $\hat{G}$ be a ring domain bounded by Jordan curves $\hat{\alpha}$*

and $\hat{\beta}$ so that $\hat{\beta}$ is in the bounded component of the complement of $\hat{\alpha}$. Let h_α be M-quasiconformal homeomorphism mapping α to $\hat{\alpha}$ and let h_β be a K-quasiconformal homeomorphism mapping β to $\hat{\beta}$. Let α be K-quasiconformal, too. Assume also that $\mathrm{mod}\, G \geq \epsilon > 0$ and

$$L^{-1} \leq \frac{\mathrm{mod}\, G}{\mathrm{mod}\, G'} \leq L \, .$$

Then there is a bound Q depending only on K, ϵ, L and M and a Q-quasiconformal homeomorphism Υ of the plane so that Υ equals h_α on the unbounded component of the complement of α and equals h_β on the bounded component of the complement of β. In addition, if the curves α and β are symmetric with respect to the real line, $h_\alpha(\bar{z}) = \overline{h_\alpha(z)}$ on α while $h_\beta(\bar{z}) = \overline{h_\beta(z)}$ on β, h_β and h_α preserve the ordering of pairs of points $\beta \cap \mathbb{R}$ and $\alpha \cap \mathbb{R}$, respectively, on the real line, then Υ can be made real on G.

Proof:

In the first step of the proof, we show that we can assume without a loss of generality that α, β, $\hat{\alpha}$ and $\hat{\beta}$ are all M-quasiconformal. This will adjust K, L, M and ϵ, but the new values will depend on the old ones only. For α M-quasiconformality was explicitly assumed and for $\hat{\alpha}$ it follows since $\hat{\alpha} = h_\alpha(\alpha)$. Let g be the Riemann mapping from the unbounded component of the complement of $\hat{\beta}$ onto the unit disk so that ∞ gets sent to 0. Inside the unit disk we get rings $g(\hat{G})$ and $g(h_\beta(G))$. Both rings separate 0 from the unit circle and both have moduli bounded from below in terms of ϵ, K and L. We claim that both rings contain a circular ring $\{z : r < |z| < 1\}$ with $r < 1$ depending only on the lower bound of their moduli. This follows from Grötzsch's module theorem (page 54 in [25]). This theorem says that if a ring separates the unit circle from points 0 and r, then its modulus is at most $\mu(r)$ where μ is a strictly decreasing function with limit 0 as $r \to 1$. Moreover, $\mu(r) < \log \frac{4}{r}$, see ibidem, page 61. Let us pick $r = \mu^{-1}((\mathrm{mod}\, G) \cdot \min\{L^{-1}, K^{-1}\})$ where the argument of μ^{-1} is clearly o lower estimate of the minimum of moduli of the rings $\hat{G}$ and $h_\beta(G)$. Since $\mathrm{mod}\, G \geq \epsilon$, then $r \leq Q_2 < 1$ where Q_2 is a function of only ϵ, K and L. Also, we have

$$\log \frac{1}{r} = \frac{\log \frac{1}{r}}{\mu(r)} \mu(r) \geq \frac{\log \frac{1}{r}}{\log 4 + \log \frac{1}{r}} \min\{L^{-1}, K^{-1}\} \mathrm{mod}\, G \qquad (5.2)$$

$$\geq \frac{\log \frac{1}{Q_2}}{\log 4 + \log \frac{1}{Q_2}} \min\{L^{-1}, K^{-1}\} \bmod G = Q_3 \bmod G$$

where Q_3 is a function of ϵ, K and L only. Now, we define $\hat{\beta}_1$ to be a preimage by g of the circle centered at 0 and the radius $\sqrt{r}$. Let $\beta_1 = h_\beta^{-1}(\hat{\beta}_1)$. The modified ring domains are G_1 bounded by α and β_1 and $\hat{G}_1$ bounded by $\hat{\alpha}$ and $\hat{\beta}_1$. The topological conditions of Lemma 5.3.1 are still satisfied by the new domains and the old functions h_α and h_β. By Lemma 4.2.1, $\hat{\beta}_1$ is quasiconformal with a bound depending on Q_2, and the bound for β_1 is at most K times larger. Moreover, the modulus of $\hat{G}_1$ is bounded as follows.

$$\frac{1}{2} \log \frac{1}{r} \leq \bmod \hat{G}_1 \leq \bmod \hat{G} \leq L \bmod G .$$

The modulus of G_1 is bounded by

$$\frac{1}{2K} \log \frac{1}{r} \leq \bmod G_1 \leq \frac{K}{2} \log \frac{1}{r} .$$

Using these estimates and (5.2) we can introduce new $\epsilon_1 = \frac{\epsilon Q_3}{2K}$ and $L_1 = 2KLQ_3$. Both new estimates depend only on ϵ, K and L so we proved our reduction. Finally, we have to consider the additional claim when the situation is symmetric with respect to the real line. In that case, Lemma 2.3.1 allows us to assume that h_β is real on the unbounded component of the complement of β. Similarly, the Riemann mapping g of a symmetric region can be chosen to be real. So our construction will result in a symmetric curve β_1 on which $\overline{h_\beta} = h_\beta(\bar{z})$. In the next step, we will use the old notations G, α, ϵ, etc.

In the second step, we further reduce the setting of Lemma 5.3.1 to the case when β and $\hat{\beta}$ are both the unit circle, while α and $\hat{\alpha}$ are also circles centered at 0. To this end, we consider the canonical mappings onto G and $\hat{G}$. We show that these canonical mappings have Q_4-quasiconformal extensions to the whole plane, called h and $\hat{h}$, respectively. To prove our claim, let us quote Theorem 6.4 from page 86 of [25]:

Fact 5.3.1 *Let D and D' be domains with free boundary curves C and C', respectively, and $w : D \to D'$ a K-quasiconformal mapping*

*which can be extended to a homeomorphism of $D \cup C$ onto $D' \cup C'$. Let
G and G' be the components of the complement of C and C' which
contain D and D' respectively. Then there exists a quasiconformal
mapping of G onto G' which has the same boundary values as w on
C and whose maximal dilatation is less than a bound depending only
on K and the domain D.*

Fact 5.3.1 can be applied to extend the canonical mapping of G
as follows. Take w equal to this canonical mapping, but restrict it
so that $D = \{z : 1 < |z| < e^\epsilon\}$. Let C be the unit circle. The
maximal dilatation bound of the extension w_1 depends only on ϵ.
The mapping w_1 sends the unbounded component of the complement
of C to the unbounded component of the complement of β, but we
can consider its quasiconformal reflection w_2 which maps between
the bounded components. The maximal dilatation of w_2 is bounded
in terms of ϵ and M. The union of w_2 and the canonical map sends
the bounded component of the complement of the outer circle to
the bounded component of the complement of α. Note that it is
quasiconformal with the same bound as w_2 because the unit circle
is a removable curve. The quasiconformal reflection of this mapping
gives us the desired h. The construction of $\hat{h}$ is quite analogous.

Then we define $h_{\alpha,1} = \hat{h}^{-1} \circ h_\beta \circ h$ and $h_{\beta,1} = \hat{h}^{-1} \circ h_\alpha \circ h$. To
complete the second step, observe that in the case of a symmetric
ring domain the canonical mapping can be chosen to be real, and
thus the mappings $h_{\alpha,1}$ and $h_{\beta,1}$ are symmetric on their respective
circles. Now we again return to the old notations like G, h_α, etc
meaning objects after the reduction in the second step.

Now unwind the rings G and $\hat{G}$ using the universal covering map
$z \to \exp(-2\pi i z)$. G is covered by the strip $\{z = x + iy : 0 < y < R\}$
and $\hat{G}$ is covered by the strip $\{z = x + iy : 0 < y < \hat{R}\}$ with
$R \geq \epsilon/2\pi$ and $L^{-1} \leq \hat{R}/R \leq L$. The mapping h_β lifts to $x \to$
$H(x)$ where $H(x)$ is Q-quasi-symmetric and Q depends on K. Also,
$H(x) = x + P(x)$ where P is 1-periodic and can be chosen so that
$P(0) \in [0,1)$. Similarly, h_α lifts to a mapping

$$x + iR \to \hat{H}(x) + i\hat{R}$$

where $\hat{H}(x) = x + \hat{P}(x)$. The properties of $\hat{H}$ and $\hat{P}$ are the same
as observed above for H and P. If h_α and h_β are symmetric on

their respective circles, then P and $\hat{P}$ are odd. We need to find a quasiconformal interpolation between H on the lower boundary of the strips, and $\hat{H}$ between the upper boundaries. Observe, that we can assume without a loss of generality that $\hat{R} = R$. Indeed, we can build an interpolating map in this setup and then compose it with an L-quasiconformal mapping $x + iy \to x + i\frac{\hat{R}}{R}y$. Similarly, if we further reduce the situation to $R = 2n$, n integer, the added dilatation will be $(R/2n)^2$ or $(R/2n)^{-2}$, whichever turns out to be greater than 1. Since $R \geq \epsilon/2\pi$, for every R we can find an n so that this added dilatation will be at most $\max\{\frac{4\pi}{\epsilon}, 2\}$. Hence we still get a bound depending on ϵ.

To extend H, let us apply the Beurling and Ahlfors' formula: (see [25], page 83) $x + iy \to u + iv$ where

$$u(x,y) = \tfrac{1}{2}\int_0^1 \ (H(x+ty) + H(x-ty))\, dt$$
$$v(x,y) = \tfrac{1}{2}\int_0^1 \ (H(x+ty) - H(x-ty))\, dt$$

This extension is quasiconformal with a bound depending on Q (see ibidem, same page). We see that $u(x,1) = x + \int_0^1 P(t)\, dt$ while $v(x,1) = 1$. So this extension preserves the line $y = 1$ and merely slides it by $\int_0^1 P(t)\, dt$. This constant is between -1 and 2. So by composing the extension with a quasiconformal "shear" we get an extension fixed on the line $y = 1$. A mirror argument shows that $\hat{H}$ can be extended from the line $y = 2$ to fix the line $y = 1$. The union of these two extensions, together with earlier reductions, gives us the desired interpolation. In the real situation, no shear is needed and the extension is real from the formula.

$\square$

5.4 Regularity of Saturated Maps

This section is devoted to the proof of Theorem 2.1. We distinguish two cases in dependence on whether f has an odd orbit or not. Let us first assume that f and $\hat{f}$ are quadratic polynomials.

Case I. In the case when f has not any odd orbits the first return map onto the fundamental inducing domain of f has only one branch.

Otherwise, an odd orbit would be easy to procure. From Sharkovski's theorem, f has orbits with infinitely different periods.

Let Φ be the first entry map from $(-1, 1)$ into $(-q, q)$. We take as a saturated map φ the restriction of Φ to D_0 and D_{-1}. The saturated map $\hat{\varphi}$ is defined analogously. We start with two well known statements which in our setting have short proofs based on Fact 5.1.4.

- The lengths of the intervals D_0 and D_{-1} are bounded away from 0 and 2 independently from f.

- The absolute value of the derivative of the right lap f_r of f on D_{-1} is bounded away from zero and infinity by a constant independent from f.

To prove the first statement we take a fixed polynomial, for example Chebyshev polynomial, which has intervals D_0 and D_{-1} nested inside $(-1, 1)$. By Fact 5.1.4, the lengths of the intervals D_0 and D_{-1} for f depends only on Q' and the lengths of the corresponding intervals for the Chebyshev polynomial. Hence, the bounds are independent from f.

By Fact 5.1.4, the mean value $|D_0|/|D_{-1}|$ of the derivative of f_r on the interval D_0 is independent from f. Monotonicity of f_r implies that f_r is $1/8|D_0|$-extendible (see Definition 2.1.1). By the real Köbe lemma, the derivative of f_r on D_{-1} is bounded from zero and infinity by a constant independent from f.

The $L(K)$-regularity given $(-1, 1), (-1, 1)$ was verified in the proof of Theorem 2.1 in the Feigenbaum case.

Case II. We are left with the situation in which odd orbits exist. In this case, we can construct type I terminal box mappings ϕ and $\hat{\phi}$, with central branches ψ and $\hat{\psi}$, central domains B and $\hat{B}$ and ranges of central branches B' and $\hat{B}'$, respectively. They also come with a Q-quasiconformal branchwise equivalence Υ where Q is a constant independent of the polynomials in question. The mapping Υ also satisfies $\Upsilon(\bar{z}) = \overline{\Upsilon(z)}$ on the boundary of the domain of ϕ and by Lemma 2.3.1 we may assume that Υ maps $\mathbb{R} \setminus (-1, 1)$ into itself in an orientation preserving fashion. Finally, the central domains of both holomorphic continuations of terminal box mappings are nested in their respective ranges with some modulus $\epsilon > 0$ again independent of the polynomials in question.

The saturated maps will be constructed as follows. On the maximal restrictive interval I of f (which is the same as the restrictive interval of the terminal mapping), φ will be the identity. On the central domain of the terminal mapping outside of the restrictive interval, φ will be undefined. The domain of every monotone branch of the terminal map contains a preimage of the restrictive interval. On each of these preimages, φ is defined and equal to the restriction of the monotone branch. This defines φ on the domain $D_0 = (-q, q)$ of f.

To prove the first claim of Theorem 2.1 it is enough to verify that the domain of φ contains the returns of the forward critical orbit of f to its fundamental inducing domain. The domain of the terminal map had this property, since all points not in its domain are non-recurrent to 0. If the domain of φ failed to contain all returns of the forward critical orbit, then the forward critical orbit would enter the central domain outside of the restrictive interval. That would mean that some image of the restrictive interval intersects the central domain. This image, call it I_m, could not intersect the boundary of the central domain, since this boundary is eventually mapped to the boundary of the fundamental inducing domain by f. But if it is contained in the central domain outside of the restrictive interval, then it cannot be mapped into itself by the central branch, since everything in this region is eventually pushed outside of the central domain.

The mapping $\hat{\varphi}$ is constructed in the same way using the dynamics of $\hat{f}$ and its terminal map.

The first two claims of Theorem 1.1 are now obvious. The third claim follows since the domains D_n satisfy

$$|D_n| \leq K \mathrm{dist}(D_n, \{-1, 1\})$$

where K is independent of f and $\hat{f}$. This is a well-known fact, and can be derived from the "universality of inner staircases" stated as Fact 5.1.4.

A modulus estimate. The small Julia set of ψ is denoted by $J(\psi)$ and defined as the boundary of $\bigcap_{i=1}^{\infty} \phi^{-i}(B')$. If f is critically recurrent than both the Julia set $J(f)$ of f and the small Julia set $J(\psi)$ do not separate the plane.

One can conjugate the central branch ψ of the terminal map on B' with a quadratic polynomial based on the Straightening Theorem, see Fact 3.1.2.

Lemma 5.4.1 *Let I be the maximal restrictive interval of a critically recurrent quadratic polynomial f. Suppose that ϕ is a terminal box mapping induced by f. Choose a topological disk D so that $I \subset D \subset B$, $\psi(D) \supset \overline{D}$ and $\mathrm{mod}\,(\psi(D) \setminus \overline{D}) \geq \epsilon_0$. Then for every ϵ_0 there is a constant K, independent of f, so that*

$$\frac{\mathrm{mod}\,(D \setminus I)}{\mathrm{mod}\,(D \setminus J(\psi))} \leq K .$$

Proof:

We apply Fact 3.1.2 to the central branch ψ of the terminal box mapping ϕ restricted to D. As the outcome we obtain a quadratic polynomial $f_a = a(1 - z^2) - 1$. This polynomial is critically recurrent since f is so. Thus $1 \leq a < 2$ and by algebra, every point z with $|z| > 1$ escapes to infinity under iterates of f_a. Therefore, diam $J(f_a) \leq 2$. Also, we have that the Julia set $J(f_a) \supset (-1, 1)$.

If H is the conjugacy, i.e. $H \circ \psi = f_a \circ H$, then we look at $H(D)$. Notice that since H quasiconformal with the bound depending only on ϵ_0, so it will be sufficient to bound

$$\frac{\mathrm{mod}\,(H(D) \setminus [-1, 1])}{\mathrm{mod}\,(H(D) \setminus J(f_a))} .$$

Distinguish two cases:

1. $H(D) \supset D(0, 10)$,

2. the opposite.

In the first case, take the Riemann mapping g from $H(B)$ onto the unit disk $D(0, 1)$, fixing 0 and sending 1 to a real value $b > 0$. By Grötzsch's module theorem (see [25], page 54),

$$\mathrm{mod}\,(H(D) \setminus [-1, 1]) \leq \mathrm{mod}\,(D(0, 1) \setminus (0, b)) = \mu(b) \leq \log \frac{4}{b} . \quad (5.3)$$

On the other hand, by Köbe's distortion lemma, $g(J(f_a))$ fits inside the disk of radius Qb where Q is a constant. So,

$$\mod (D(0,1) \setminus g(J(f_a))) \geq \log \frac{1}{Qb} \,.$$

This implies the desired bound as long as $b \leq \frac{1}{Q^2}$. Otherwise, from (5.3) we have

$$\mod (H(D) \setminus [-1,1]) \leq 2 \log(2Q)$$

while

$$\mod (D(0,10) \setminus J(f_a)) \geq \log 10 \,.$$

In the second case, by Teichmüller's module theorem (see [25], page 56), the modulus of $H(D) \setminus [-1,1]$ is bounded from above by a constant. On the other hand, $\mod (H(D) \setminus J(f_a))$ is bounded from below in terms of ϵ_0.

This ends the proof of Lemma 5.4.1.

□

Lemma 5.4.2 *Let f and $\hat{f}$ be two critically recurrent unimodal polynomials which induce terminal box mappings ϕ and $\hat{\phi}$ respectively. Let Q be the maximal dilatation of the branchwise equivalence Υ. Then one can find a Q-quasiconformal homeomorphism H which maps the annulus $B \setminus J(\psi)$ onto $\hat{B} \setminus J(\hat{\psi})$ and additionally for every m positive, sends the curve $\psi^{-m}(B)$ onto $\hat{\psi}^{-m}(\hat{B})$.*

Proof:
This homeomorphism is constructed from the branchwise equivalence Υ between the terminal maps. Once we restrict the domain of definition of Υ to the complement of B, we find Υ_n for every n so that

$$\Upsilon_n \circ \psi^n = \hat{\psi}^n \circ \Upsilon_n \,.$$

They will match continuously to give the desired map H, which is Q-quasi-conformal.

□

Regularity of branchwise equivalences. To conclude Theorem 2.1, it remains to prove that the pair $(\varphi, \hat{\varphi})$ is $L(K)$-regular given $(-1, 1), (-1, 1)$. Suppose that an order-preserving K-quasisymmetric homeomorphism υ has been given between the restrictive intervals of f and $\hat{f}$. In order to construct the saturated branchwise equivalence, the rough idea would be to interpolate between Υ on the complement of B and υ on the restrictive interval by some quasiconformal map and then propagate it by pull-back on the domains of all univalent branches. The tool here would be Lemma 5.3.1. There are some technical issues that need to be taken care of.

Step I.   *Observe that Υ commutes with complex conjugation.*

If we go back to the initial branchwise equivalence constructed in Theorem 1.3, we see that it commutes with complex conjugation. Since our operations in constructing Υ consisted of pull-back by real polynomials and corrections on symmetric domains, it is clear that they kept the maps symmetric.

Step II.   Pick the Jordan curve α which splits the annulus $B \setminus \overline{\psi^{-1}(B)}$ into two annuli of modulus half as big. This curve is Q_1-quasiconformal where Q_1 is a constant by Lemma 4.2.1. Let D be bounded connected component of the complement of α, and $\hat{D} := \Upsilon(D)$.

Step III.   We construct a quasiconformal mapping Υ_1 between Jordan domains $U \subset D$ and $\hat{U} \subset \hat{D}$ which does the following:

- maps $B \cap \mathbb{R}$ onto $\hat{B} \cap \mathbb{R}$,

- coincides with υ on the restrictive interval,

- the ratio $\bmod(D\backslash U)$ to $\bmod(\hat{D}\backslash\hat{U})$ is bounded from above and below by positive constants independent of the polynomials.

To construct Υ_1, first build a real Q_2-quasiconformal homeomorphism Υ_2 of the plane which extends υ. This is easy to do since υ can first be extended to a homeomorphism of the entire line with a bounded loss of quasi-symmetric estimate (see Lemma 7.1, page 89 of [25]) then continued to the plane by the Beurling-Ahlfors theorem, see ibidem, Theorem 6.3 on page 83. The constant Q_2 is independent of the polynomials in question. Now, if we pick a very small Jordan domain U around I then its image by Υ_2 is also very close to $\hat{I}$ and, by the continuity of the module ([25], Lemma 6.4), the moduli $\mathrm{mod}\,(D \setminus U)$ and $\mathrm{mod}\,(\hat{D} \setminus \hat{U})$ can be made arbitrarily close to $\mathrm{mod}\,(D \setminus I)$ and $\mathrm{mod}\,(\hat{D} \setminus \hat{I})$, respectively. Now the last item follows from Fact 3.1.2 and Lemma 5.4.2.

Now we are in the position to use Lemma 5.3.1 to interpolate between Υ_1 on $\beta := \partial U$ and Υ on the complement of D. Let A be an annulus with boundary $\alpha \cup \beta$ and $\hat{A}$ an annulus with boundary $\Upsilon(\alpha) \cup \Upsilon_1(\beta)$. The ratio of the moduli of these two annuli is bounded away from zero and infinity by a constant independent from f and $\hat{f}$. Lemma 5.3.1 will give the needed interpolation, commuting with the complex conjugation, between these maps, with maximal dilatation bounded depending only on K. Call this map Υ_2.

Now we take any univalent branch of the terminal box mapping and pull back Υ_2. The mappings match continuously since $\Upsilon_2 = \Upsilon$ except on the central domain. The restriction h to the real axis of the mapping obtained in this way is an $L(K)$-quasi-symmetric branchwise equivalence between the saturated maps φ and $\hat{\varphi}$ given $((-1,1),(-1,1),v)$. Moreover, h outside of $(-1,1)$ is the restriction of Υ to the real line and thus does not depend on v. We conclude that the pair $(\varphi,\hat{\varphi})$ is $L(K)$-regular as needed. This completes the proof of Theorem 2.1.

5.5 Straightening Theorem

We will now prove Fact 3.1.2.

First, we may assume that the boundary of V, further called α, and the boundary of U, further called β, are L-quasiconformal Jordan curves where L depends only on K. At the same time, we have to weaken the requirement that both the critical value of f and its image are in U, replacing it with the hypothesis that just the critical value is in U. Indeed, let H be a uniformizing map from the annulus $V \setminus \overline{U}$ onto a round ring domain. Take α as the preimage by H of the circle that splits this round annulus into two annuli of equal modulus, and make β the preimage of α by f. Their quasiconformality follows from Fact 4.2.3 with T equal to the round annulus. Also, f now extends analytically to a neighborhood of α.

Next, without loss of generality, assume that α is a circle centered at 0 with radius $\rho := \exp(2\mathrm{mod}\,(V \setminus \overline{U}))$ and 0 is the critical point of f. This is achieved by changing the coordinates by the Riemann map of V. Notice that β becomes a Jordan curve which contains 0 in the bounded connected component of its complement and is still separated from α by an annulus with modulus at least K. It follows that the Riemann mapping had distortion on β bounded in terms of K, and β remains quasiconformal with distortion dependent on K only.

Now consider a pair of branched covers of a $D(0, \rho + \eta)$, $\eta > 0$, given by f and $Q(z) = z^2$. Construct a homeomorphism h from $D(0, \rho + \eta)$ onto itself which is the identity outside of $D(0, \rho)$, is symmetric with respect to the real line and satisfies $h(0) = f(0)$. The critical value $f(0)$ is surrounded by β and its distance to the circle $C(0, \rho)$ is bounded from below by a positive constant dependent on K. Hence, by Lemma 5.2.3, h can be made L-quasiconformal with L depending on K only. Let h' be a lifting given by $h \circ Q = f \circ h'$. Now apply Lemma 5.3.1 to interpolate between the identity outside of $C(0, \rho)$ and h' in the disk bounded by $C(0, \sqrt{\rho})$. What results is an $L(K)$-quasiconformal homeomorphism H of the whole plane. The mapping $\Gamma := H \circ Q \circ H^{-1}$ equals f on U and Q outside of the region bounded by $C(0, \rho)$. Clearly, Γ is an $L(K)$-quasiregular degree 2 branched cover of the sphere, and is conformal except on a fundamental region $V \setminus \overline{U}$.

From here the proof proceeds as in the original paper [8]. We construct an invariant measurable conformal structure ν by setting $\nu = dz$ outside of $\overline{V}$ and then pulling it back by iterations of Γ. For points whose orbits by f never leave V, we set $\nu = dz$. This gives a measurable conformal structure $\nu = a(z)d\overline{z} + b(z)dz$ which is invariant by Γ and

$$|a(z)/b(z)| \le \frac{L(K) - 1}{L(K) + 1}$$

almost everywhere. In addition, $a(\overline{z}) = \overline{a(z)}$ and $b(\overline{z}) = \overline{b(z)}$. By the measurable Riemann mapping theorem, see [3], we find a quasi-conformal homeomorphism P of the sphere fixing 0, 1 and ∞, with $P^*(dz) = \nu$. By the symmetry of ν, $\overline{P(\overline{z})} = P(z)$. Then $P \circ \Gamma \circ P^{-1}$

is a degree 2 holomorphic branched cover of the sphere fixing the infinity, hence a quadratic polynomial.

The straightening theorem has been proved with $H := H^{-1} \circ P^{-1}$.

Bibliography

[1] Bers, L.: *On moduli of Kleinian groups*, Russ. Math. Surveys **29:2** (1974), pp. 86-102

[2] Bers, L. & Royden, H.L.: *Families of holomorphic injections*, Acta Math. **157**, (1986), pp. 259-286

[3] Bojarski, B.: *Generalized solutions of a system of first order differential equations of elliptic type with discontinuous coefficients* (in Russian), Mat. Sbornik **43** (1957), pp. 451-503; see also: Ahlfors, L. & Bers, L.: *Riemann's mapping theorem for variable metrics*, Acta Math. **72** (1960), pp. 385-404

[4] Branner, B. & Hubbard, J.H.: *The iteration of cubic polynomials, Part II: Patterns and parapatterns*, Acta Math. **169**, 229-325 (1992)

[5] Carleson, L. & Gamelin, T.: *Complex dynamics*, Springer-Verlag, New York-Berlin-Heidelberg (1995)

[6] Douady, A.: *Systemès Dynamiques Holomorphes*, Astérisque **105** (1982), pp. 39-63

[7] Douady, A. & Hubbard, J. H.: *Étude Dynamique des polynômes quadratiques complexes*, Publications Mathématiques d'Orsay no. **84-02** and **85-04** (1984)

[8] Douady, A. & Hubbard, J.H.: *On the dynamics of polynomial-like mappings*, Ann. Sci. École Norm. Sup. (Paris) **18** (1985), pp. 287-343

[9] Farkas, H. & Kra, I.: *Riemann surfaces*, Springer-Verlag, New York-Heidelberg-Berlin (1980)

[10] Fatou, P.: *Sur les equations fonctionnelles, II*, Bull. Soc. Math. France, **48** (1920), pp. 33-94

[11] Gehring, F. : *Characteristic properties of quasidisks*, Les Presses de l'Université de Montréal, Montreal (1982)

[12] Graczyk, J. & Smirnov, S.: *Fibonacci Julia sets, Conformal Measures and Hausdorff dimension*, manuscript (1995), pp. 1-22

[13] Graczyk, J. & Świątek, G.: *Induced expansion for quadratic polynomials*, Ann. Sci. École Norm. Sup. **29** (1996), pp. 399-482

[14] Graczyk, J. & Świątek, G.: *Holomorphic Box Mappings*, preprint IHES no. **IHES/M/96/76**, to appear in Astérisque.

[15] Graczyk, J. & Świątek, G.: *Polynomial-like property for real quadratic polynomials*, Topology Proceedings, **26** (1996), pp. 33-112

[16] Graczyk, J. & Świątek, G.: *Generic hyperbolicity in the logistic family*, Ann. of Math. **146** (1997), pp. 1-52

[17] Guckenheimer, J.: *Sensitive dependence on initial conditions for one dimensional maps*, Commun. Math. Phys. **70** (1979), pp. 133-160

[18] Guckenheimer, J.: *Limit sets of S-unimodal maps with zero entropy*, Commun. Math. Phys. **110** (1987), pp. 655-659

[19] Heckman, Ch.: *Ph.D. thesis*, Stony Brook (1996)

[20] Jakobson, M.: *Absolutely continuous invariant measures for one-parameter families of one-dimensional maps*, Commun. Math. Phys. **81** (1981), pp. 39-88

[21] Jakobson. M.: *On the boundaries of some domains of normality for rational maps*, report Institut Mittag-Leffler, **15**, 1983

[22] Jakobson, M.: *Quasisymmetric conjugacies for some one-dimensional maps inducing expansion*, Contemporary Mathematics **135**, (1992), pp. 203-211

[23] Jakobson, M. & Świątek, G.: *Metric properties of non-renormalizable S-unimodal maps*, Ergod. Th. and Dynam. Sys. **14** (1994), pp. 721-755

[24] Jakobson, M. & Świątek, G.: *Metric properties of non-renormalizable S-unimodal maps: II. Quasisymmetric conjugacy classes*, Ergod. Th. & Dynam. Sys. **15** (1995), pp. 871-938

[25] Lehto, O. & Virtanen, K.: *Quasikonforme Abbildungen*, Springer-Verlag, Berlin-Heidelberg-New York (1965)

[26] Levin, G & Van Strien, S.: *Real polynomials with locally connected Julia sets*, to appear in Ann. of Math.

[27] Mañe, R. , Sad, P. & Sullivan, D.: *On the dynamics of rational maps*, Ann. Sci. École Norm. Sup. **16** (1983), pp. 193-217

[28] Martens, M.: *Distortion results and invariant Cantor sets of unimodal mappings*, Erg. Th. & Dyn. Sys. **14** (1994), pp. 331-349

[29] McMullen, C.: *Complex dynamics and renormalization*, Annals of Mathematics Studies **135**, Princeton University Press, Princeton (1994)

[30] de Melo, W. & van Strien, S.: *One-dimensional dynamics*, Springer-Verlag, New York (1993)

[31] Milnor, J.: *The Yoccoz theorem on local connectivity of Julia sets. A proof with pictures.*, class notes, Stony Brook, (1991-92)

[32] Milnor, J. & Thurston, W.: *On iterated maps of the interval I, II*, preprint (1977), in: Lecture Notes in Mathematics, Vol. 1342 (Springer, Berlin, 1988)

[33] Nowicki, T. & Sands, D.: *Non-uniform hyperbolicity and universal bounds for S-unimodal maps*, to appear in Inventiones

[34] Perez-Marco, R.: *Topology of Julia sets and hedgehogs*, preprint Université de Paris-Sud no. **94-48**

[35] Przytycki, F.: *Iterations of holomorphic Collet-Eckmann maps: Conformal and invariant measures. Appendix: on non-renormalizable quadratic polynomials*, Trans. Amer. Math. Soc. **350** (1998), pp. 717-742

[36] Sands, D.: *Complex bounds*, manuscript of 1996

[37] Slodkowski, Z.: *Holomorphic motions and polynomial hulls*, Proc. AMS, **111**, pp. 347-355

[38] Sullivan, D.: *Quasiconformal homeomorphisms and dynamics I: a solution of Fatou-Julia problem on wandering domains*, Ann. Math. **122** (1985), pp. 401-418

[39] Sullivan D.: *Bounds, quadratic differentials and renormalization conjectures*, in: *Mathematics into the Twenty-First Century*, AMS Centennial Publications (1991)

[40] Teichmüller, O.: *Untersuchungen über konforme und quasikonforme Abbildung*, Deutsche Mathematik **3**, (1938), pp. 621-678

[41] Yoccoz, J.-C.: unpublished results, see the discussion in [31]

Index

Beltrami equation, 7
Beltrami bound, 121
Beurling-Alhfors extension, 36
box mapping ϕ, 14
    Jordan, 14
    symmetric, 69
    terminal, 15
    type I, 15
    type II, 14
Böttker coordinate, 110
bounded turning, 41
branches, 14
    central ψ, 14
    immediate, independent,
        maximal, 69
    combinatorial structure, 68
    parent, 68, 69
    postcritical, 76
    subordinate, 69
branchwise equivalence, 19, 31
    correction, 124, 138
    similar, 21
    quasiconformal, 18

canonical mapping, 72
central domain B, 14
Chebyshev polynomial, 115, 134
close return, 15
complex bounds, 45
cost of correcting, 124

critical filling, 15
critically consistent, 22
critically recurrent, 10
conformal roughness, 88
    quasi-invariance, 95
cross-ratios, 53, 54
    contracting **Cr**, 54
    expanding **Poin**, 53
cutting time, 47
    proper, 47

dense hyperbolicity, 6
distortion, 26, 29
double rays, 110
dynamical extension, 69

ϵ-extendible, 26
equipotential curves, 110
equicontinuity, 103
escaping time, 15

Fatou, 4, 9
Feigenbaum map, 25, 34
Fibonacci returns, 106
filling-in, 15
first entry time, 12
first return time, 13
fixed point π_n, 50
fixed point χ_n, 50

geodesic neighborhood, 28